The Lipoprotein Molecule

NATO ADVANCED STUDY INSTITUTES SERIES

A series of edited volumes comprising multifaceted studies of contemporary scientific issues by some of the best scientific minds in the world, assembled in cooperation with NATO Scientific Affairs Division.

Series A: Life Sciences

Recent Volumes in this Series

Volume 8 – Meat Animals: Growth and Productivity
edited by D. Lister, D. N. Rhodes, V. R. Fowler, and M. F. Fuller

Volume 9 – Eukaryotic Cell Function and Growth: Regulation by Intracellular Cyclic Nucleotides
edited by Jacques E. Dumont, Barry L. Brown, and Nicholas J. Marshall

Volume 10 – Specificity in Plant Diseases
edited by R. K. S. Wood and A. Graniti

Volume 11 – Surface Membrane Receptors: Interface Between Cells and Their Environment
edited by Ralph A. Bradshaw, William A. Frazier, Ronald C. Merrell, David I. Gottlieb, and Ruth A. Hogue-Angeletti

Volume 12 – Nucleic Acids and Protein Synthesis in Plants
edited by L. Bogorad and J. H. Weil

Volume 13 – Prostaglandins and Thromboxanes
edited by F. Berti, B. Samuelsson, and G. P. Velo

Volume 14 – Major Patterns in Vertebrate Evolution
edited by Max K. Hecht, Peter C. Goody, and Bessie M. Hecht

Volume 15 – The Lipoprotein Molecule
edited by Hubert Peeters

Volume 16 – Amino Acids as Chemical Transmitters
edited by Frode Fonnum

The series is published by an international board of publishers in conjunction with NATO Scientific Affairs Division

A	**Life Sciences**	Plenum Publishing Corporation
B	**Physics**	New York and London
C	**Mathematical and Physical Sciences**	D. Reidel Publishing Company Dordrecht and Boston
D	**Behavioral and Social Sciences**	Sijthoff International Publishing Company Leiden
E	**Applied Sciences**	Noordhoff International Publishing Leiden

The Lipoprotein Molecule

Edited by
Hubert Peeters
Lipid and Protein Department LBS
Brussels, Belgium

PLENUM PRESS • NEW YORK AND LONDON
Published in cooperation with NATO Scientific Affairs Division

Library of Congress Cataloging in Publication Data

Nato Advanced Study Institute on the Lipoprotein Molecule, Bruges, Belgium, 1977
The lipoprotein molecule.

(NATO advanced study institutes series: Series A, Life sciences; v. 15)
Includes index.
1. Blood lipoproteins–Congresses. I. Peeters, Hubert, 1919- II. Title. III. Series.
QP99.3.L52N37 1977 612'.0157 78-2388
ISBN-13: 978-1-4684-2783-7 e-ISBN-13: 978-1-4684-2781-3
DOI: 10.1007/978-1-4684-2781-3

Proceedings of the NATO Advanced Study Institute on the Lipoprotein
Molecule held in Bruges, Belgium, May 8–20, 1977

A Division of Plenum Publishing Corporation
227 West 17th Street, New York, N.Y. 10011

Preface

The amount of recent information collected about the molecular composition, structure, and function of the plasma lipoproteins, in man as well as in experimental animals, is very large. In this volume an assessment of our knowledge as it stands today has been organized within a framework of four general topics, the first about analytical data, the second about structure, the third about metabolism, and the last about molecular variation and pathology. Thus the analytical, theoretical, experimental, and applied aspects of the topic have been treated in conjunction with each other.

The analytical data in human and nonhuman primates were obtained after ultracentrifugal or electrophoretic separation of the protein class from the native serum. Comparisons of various methods were not forgotten. The main categories of lipoproteins, namely High Density (HDL), Low Density (LDL), and Very Low Density (VLDL), are treated separately, and lipid moiety and the polypeptide moiety are carefully described.

In the theoretical field of reassembly of lipoprotein particles by relipidizing delipidated polypeptides, the structural aspects of lipoprotein and the stabilizing effect of phospholipids on the polypeptide structure were studied. The opposite direction of the process, namely degradation of a lipoprotein by means of lipolytic enzymes, has also been investigated and described.

Some inborn errors of metabolism and some acquired disorders such as jaundice and also the pathology of atherosclerosis have been related to peculiar and often specific lipoprotein patterns. These in turn can be modified by diet or drugs, and the efforts to restore normality are largely based on the analytical and structural knowledge of lipoproteins described in the earlier sections of the course.

Discussions, round tables, and short spontaneous sessions organized by participants themselves during the course proved to teachers and students alike that this initiative was successful.

We are grateful to the faculty members for preparing their manuscripts carefully and also to Plenum Press for the speedy publication of this volume.

Hubert Peeters

Contents

Introductory Remarks

THE CHALLENGE OF THE LIPOPROTEIN MOLECULE

Hubert Peeters

Lipid and Protein Department, LBS

B-1180 Brussels, Belgium

The primaeval outline of a lipoprotein appeared in the course of 1926 and 27 when Macheboeuf presented a few short papers on the phospholipids of plasmaproteins. In 1928 he defended his doctoral thesis: Recherches sur les stérols, les lipides et les protéides du sérum et du plasma sanguin", Thèse pour le Doctorat ès Sciences, Paris. This NATO Advanced Study Institute thus coïncides with the 50th anniversary of the fundamental work from which the modern concept of lipoproteins emerged.

As a starting point he used his observation that lipids - not soluble in water - were solubilized in plasma through some of the plasma proteins. For these lipid-binding proteins he coined a new terminology - no longer in use today - le "cénapse lipoprotéique". This lipid-protein bond was not easy to disrupt as ether was unable to extract all lipids from the serum. These observations were done on horse serum which was readily available at the Pasteur Institute.

He counterchecked his hypothesis of the existence of peculiar lipid-protein structures in serum or plasma, his so called "cénapses", by mixing protein with fatty acids in the presence of bile salts, thus reconstituting some sort of a plasma lipoprotein. In 1931 he applied his views to lipoid-nephrosis - the nephrotic syndrome - because he knew, about the abnormal amounts of fat present in the plasma of these patients.

Thus the work of Macheboeuf contains the fundamental contribution to the lipoprotein concept which is the topic of this course. How did this first image evolve?

Density was applied as a separating factor for the intricate mixture of proteins which were suspected in plasma and the intriguing problem of the X protein was solved by the isolation of lipoproteins by the ultracentrifugal flotation procedure. The presence of the Donner Laboratories at this meeting shows the sustained interest for centrifugation over the bias of microequipment and the importance of the density parameter in the establishment of lipoprotein density classes.

Free zone electrophoretic patterns were seriously disturbed by lipoprotein turbidity, but on paper special lipid staining techniques helped us to relate the ultracentrifugal to the mobility patterns. In due turn immunology played its part in distributing the apopolypeptides into lipoprotein families and we are awaiting better reagents and methodology, in order to apply this knowledge on a practical clinical scale.

A morphologic approach through electronmicroscopy confirmed the ultracentrifugal concept of lipoprotein classes at the molecular or rather at the particle scale and also helped to detect anomalies connected with enzyme deficiencies. Some intriguing physicochemical problems related to the appearance of discshaped and spherical particles are being solved by means of elegant physical methods such as X-Ray scattering, NMR and several others.

However not one of these techniques has yet solved the area of uncertainty which floats around the identity of a lipoprotein molecule. It is not clear, neither from the density-, morphology-, mobility-, nor immunology parameters, that lipoproteins are precisely built. On the contrary, an aura of vagueness always prevails. The same polypeptides are present in almost every lipoprotein although in a different amount and proportions.

The chemical composition of the lipid moieties became a fruitful approach towards the identity of a lipoprotein. Once density and mobility were related to the lipid load, the distribution of the lipid classes with their many variants as well in their alcoholic as in their acyl moieties - either ester or ether - became a clue to the identity, function and metabolism of the molecule or particle.

Despite the analytical imprecision of the methods in use in lipid analysis and partly due to the problems related with the isolation of precisely defined lipoproteins, an increasing amount of data is appearing which tends to stress the rigour of the lipoprotein lipid load. Could one expect that lipids would be carried as bulk material on the delicate spine of a polypeptide? Ongoing research points towards the precise location of precise phospholipids on the polypeptide backbone. The precision of the lipid load probably reaches as far as the double bonds. It is not clear today to what extent the lipid load is determined by the polypeptide end of the molecule and how far lipids are allowed to interexchange inside and between their own subgroups. A question of this order is the distribution of a given load of unsaturated double bonds over the fatty acids of a lipoprotein molecule. We tend to believe that the number of double bonds could partly, yet imperfectly, substitute for each other to a certain extent, e.g. in such a way that 20 double unsaturated fatty acids could be replaced by 40 monounsaturated ones. If such type of reasoning were correct the amount of lipid in a partcile would automatically and of necessity be greater when more unsaturated fatty acids are present.

There are more of these questions. Are permutations, inside the lipid load, allowed all the way through or are there only some of the double bonds available for such an exchange inside a given lipid class? How much variation between phospholipid families is tolerated within the same molecule? Which and how much polypeptides are required if a different lipid load is to be carried? Comparative biochemistry may bring a clue to some of these problems and should fit in with the puzzle of human pathology. Genetic disorders in man could also help to break the code in this field.

An aspect of lipoproteins which helps to sustain their fascination is their ability to pick up lipid soluble dyes which are readily observed on the colourless holoprotein background. Let us forget for a moment the quantitation problems of our staining procedures and extend the idea of stainability to any chemical one can introduce into the lipid moiety. Recently fluorescent markers have been introduced and show to behave differently in a stoggy cholesterol milieu and more loosely in a triglyceride environment. The depolarization of fluorescence is a physical parameter to be measured on intact and native lipoproteins without requiring their isolation from a mixture. Such measurements are an easy means of quality control of material under separation, under delipidation or reassembly. Along a different line this method opens the way for the study of membrane lipoproteins which might become a topic for a later course.

Evidently the metabolic aspects of lipoproteins have been extensively studied. Ever since chyle was observed in antiquity and used as an omen in the prediction of the future after the sacrifice of poultry in honour of the ancestors, some relation between lipid load and food intake had been established. The best separation methods and refined biochemical systems are required to separate the enzyme systems responsible for lipid turnover across the lipoprotein vehicle. Progress in the clinical field is largely based on a modern reading of lipid disorders of the plasma of patients and rather reminiscent of the work of the augures in antiquity. Practical enzymatic procedures for lipolytic enzymes are still to be further developed and standardized.

As a result of our insufficient knowledge the drug and diet

therapy of lipoprotein disorders is still in its infancy. We can expect a breakthrough only when the basic research will have clarified the position of plasma lipoproteins as a parameter of arterial pollution.

In this introduction we have been avoiding names, figures, graphs and tables. Thus those who are to follow can use the weapons and the bullets of their choice. It is the role of a general to decide on the line of attack, to draw a map, to describe the position, but to avoid the details.

If the faculty has been selected for its competence the course has been lined out for the sake of the students. If lipoproteins indeed cause turbidity in our biological fluids, they should generate clear physiological concepts in our minds and this was the purpose of this Advanced Nato Study Institute and is the aim of these proceedings.

Isolation and Composition of the Lipoprotein Molecule

THE PROGNOSIS FOR QUANTITATIVE LIPOPROTEIN ELECTROPHORESIS

Frank T. Lindgren

Donner Laboratory, Lawrence Berkeley Laboratory

University of California, Berkeley, California

LIPOPROTEIN vs. CHOLESTEROL AND TRIGLYCERIDE DETERMINATIONS

Surely any system for classifying lipid and lipoprotein disorders will have limitations, and this is certainly true for the Fredrickson, Levy and Lees "typing" system (1). Although during the past decade this "typing" system has allowed classification of most lipoprotein disorders into six types, there has been some dissatisfaction and disenchantment in its usage. Sometime ago, two editorials (2, 3) have advocated discontinuance of "routine electrophoresis" in clinical testing. Unfortunately, this would potentially have the effect of removing a cornerstone of the typing system, namely, the direct estimation or quantification of the major lipoprotein classes.

The real limitations of the typing system as practiced by clinical laboratories were twofold: a) it failed to categorize lipoprotein abnormalities on a genetic basis, and b) it did not provide quantitative lipoprotein data. It is ironic that at just the time when the earlier crude techniques of paper electrophoresis are being replaced by refined techniques capable of yielding such quantitative data (4, 5) some investigators are recommending wholesale abandonment of electrophoresis.

One author (2) advocates discontinuance of routine lipoprotein electrophoresis, but at the same time wishes "to try to better understand lipoprotein metabolism". This goal seems incompatible with the elimination of electrophoresis, since it would appear necessary to study both qualitatively and quantitatively all of the lipoprotein classes rather than just plasma triglyceride (TG) and cholesterol (TC) levels. The author further justifies his position by saying, "There is virtually a total overlap of information between plasma levels of cholesterol-triglycerides and lipoprotein electrophoresis (exceptions: triglyceride > 350 mg/100 ml)". I believe this is an unfortunate overstatement. The assertion that "one could quite accurately predict the lipoprotein pattern" knowing the plasma TG and TC as well as the "appearance" of the fasting plasma is probably correct in most cases, particularly if one is an experienced clinician. Yet, there are important exceptions in both normal and hyperlipoproteinemic sera. A complication is introduced by the fact that there is a similar ratio of TG/TC in both low density lipoproteins (LDL) and high density lipoproteins (HDL). The former molecules are atherogenic while there is evidence to suggest the latter are potentially helpful for necessary cholesterol clearance from tissues (6). Indeed, there is evidence to suggest that HDL is inversely related to CHD prevalence and that this relationship is largely independent of total cholesterol and LDL levels (7). An elevation of either 100 mg/100 ml LDL or 155 mg/100 ml HDL would raise both the serum cholesterol by 33 mg/100 ml. In such a case, it would be impossible to distinguish by lipid values alone, whether this elevation was due to LDL or HDL. However, this could be determined by internally standardized quantitative agarose gel electrophoresis.

The implications are twofold. At a relatively moderate plasma cholesterol level (e.g., 210-240 mg/100 ml) a physician might overlook a patient (false normal) with an elevated LDL if the HDL is abnormally low. Conversely, with a somewhat elevated plasma cholesterol, say 240-280, a patient with an unusually high HDL might be unnecessarily subjected to diet and/or a drug such as cholestyramine, with no apparent therapeutic effect. Perhaps as many as 10% of individuals screened would fall in the above categories and thus be misdiagnosed.

Another more important example of the usefulness of electrophoresis would be the discrimination of most type II b's, III's, and some IV's, one from another. This might involve some 30% of all patients with lipoprotein abnormalities (8). Except for the type III, which in addition to electrophoresis would require ultracentrifugation for confirmation, quantitative electrophoresis would permit a more refined definition of the patient's lipoprotein abnormality. Although occasional "false positives" may be identified by electrophoresis (9) of the very low density lipoprotein (VLDL) fraction, therapy for a bonafide type V would not be significantly different. Since sound medical practice involves screening members of a type III's family with electrophoretic definition of type III, some families may be screened unnecessarily.

It is agreed that it is possible to treat hyperlipoproteinemia empirically on a hit and miss basis usually with success knowing only the TG, TC and chylomicron levels. However, typing in many cases will allow a more systematic initial therapy (10) with an anticipated earlier response as well as more accurate prognosis for the patient.

It also should be conceded that the typing system has failed to put all lipoprotein abnormalities neatly into six discrete genetic types. Moreover, under dietary and/or drug manipulation an individual patient may exhibit more than one "type" at different times. Also, there are degrees of severity of each type as well as combinations of abnormalities, and these observations cannot be adequately expressed by qualitative electrophoresis and plasma TG and TC levels. What is fundamentally needed is the quantitative levels of the major lipoprotein classes to provide an appropriate measure of the "degree of abnormality"; for example, the degree of II'ness or of IV'ness. In addition, such a system could readily handle and quantitatively characterize "mixed hyperlipoproteinemia" expressing, for example, both II'ness and IV'ness. Further, these abnormalities need to quantitatively relate to established normal population values for each lipoprotein class (11). For example, the percentile rank and standard deviation score value of each lipoprotein class can appropriately identify the degree of lipoprotein abnormality as well as the degree of II'ness, IV'ness, etc. Thus, quantitative internally standardized lipoprotein electrophoresis providing this information can add a needed new dimension to the typing system. However, its potential usefulness can only be determined after a reasonable period of application and usage.

ADVANTAGES OF AN AUTOMATED MICROELECTROPHORESIS SYSTEM

Tedious manipulations and calculations have been a major obstacle in the clinical application of quantitative lipoprotein electrophoresis. To overcome these difficulties an automatic system for the microdensitometry and calculations as part of a quantitative agarose gel electrophoresis system has been developed (12). Quantitative results are internally standardized by serum cholesterol measurements using the mean wt % content of cholesterol in each major lipoprotein class (5). The hardware, described in detail elsewhere (12), consists of microdensitometer, an analog to digital converter, a cathode ray tube terminal, a teleprinter and a PDP 8/e computer equipped with a disc. The complete system is shown schematically in Fig. 1. Scans are made of electrophoretic patterns of serum alone or in combination with the 1.006 g/ml VLDL top and/or bottom preparative ultracentrifuge lipoprotein fractions (13). The analysis corrects for baseline drifts, pre-beta asymmetry and will properly identify and quantify the amount of VLDL, LDL and HDL with corrections for "sinking pre-beta" and "floating beta" in LDL and VLDL, respectively. The latter corrections require the use of the ultracentrifuge fractions. Final results are given in mg/100 ml as well as percentile and standard deviation

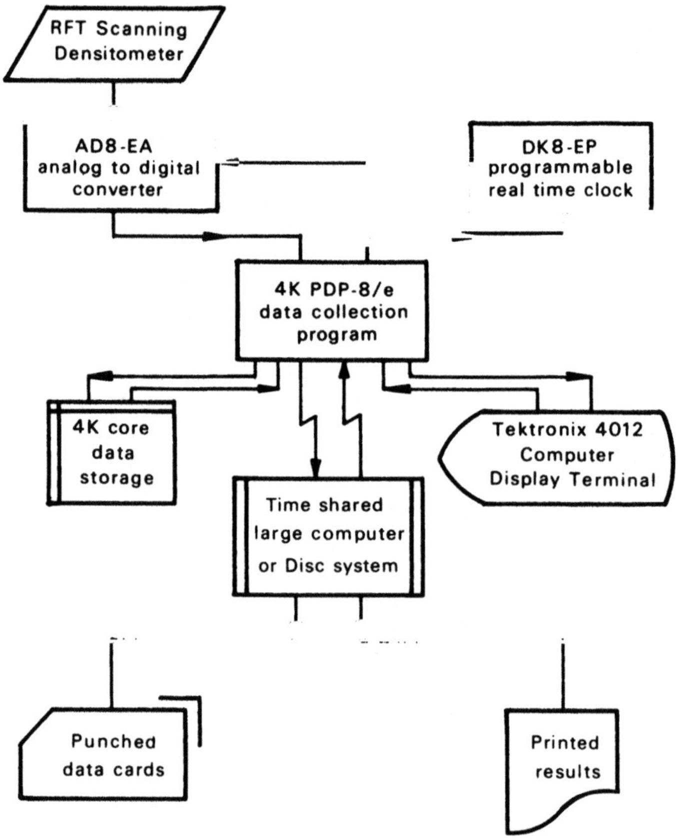

Figure 1: Schematic diagram showing the functional units and the flow of data in the automated system for quantitative lipoprotein electrophoresis. Reprinted from reference 12 by permission of the Journal of Laboratory and Clinical Medicine.

score rank of each lipoprotein class as compared to an appropriate normal reference population (11). Examples of such data with ranking for two subjects with similar lipid levels are given in Fig. 2. Such data are in a form more meaningful to the physician and patient and may provide a quantitative dimension to lipoprotein phenotyping.

The above procedure inclusive of all calculations and appropriate ranking requires about 5 minutes technician time. Further, it is anticipated that in routine screening only a small fraction, say 10%, of the samples will require the complete procedure involving lipoprotein fraction scans. The plasma scan alone will be sufficient in most cases for lipoprotein quantification and ranking. Given such conditions, two technicians could process some 250 serum samples per week of which perhaps 25-36 might involve preparative

CASE 7184 SLIDE 350 AGE 48 SEX F

G 900

CLASS	SU	MGPCT	ANUC	VS 27-45 FEMALES RANK	S.D.	46-65 FEMALES RANK	S.D.	ALL FEMALES RANK	S.D.
HDL	14983	574.4	(537.5)	99%	2.31	97%	1.85	98%	2.04
VLDL	5255	94.4	(74.9)	84%	1.01	34%	-.40	45%	-.12
LDL	24692	540.9	(481.8)	97%	1.94	80%	.86	89%	1.24
TG		108.0		95%	1.64	38%	-.30	51%	.02
CS		289.0		100%	4.07	91%	1.34	97%	1.93

FRACTIONS UTILIZED WS ONLY STANDARDIZED USING CS ONLY
ALPHA STAINING FACTOR= 1.75

CASE 7188 SLIDE 345 AGE 45 SEX F

G 900

CLASS	SU	MGPCT	ANUC	VS 27-45 FEMALES RANK	S.D.	46-65 FEMALES RANK	S.D.	ALL FEMALES RANK	S.D.
HDL	3522	178.5	(227.1)	10%	-1.30	5%	-1.61	8%	-1.43
VLDL	3946	93.7	(74.3)	84%	.99	34%	-.41	45%	-.12
LDL	25919	750.5	(666.7)	100%	4.47	100%	2.72	100%	3.26
TG		111.0		96%	1.79	39%	-.27	52%	.06
CS		292.0		100%	4.20	92%	1.41	98%	2.00

FRACTIONS UTILIZED WS ONLY STANDARDIZED USING CS ONLY
ALPHA STAINING FACTOR= 1.75

Figure 2: Teleprinter output format with ranking for cases 7184 and 7188. Note that with similar TC and TG levels the HDL and LDL levels are substantially different. Abbreviations are: SU, scanning units; WS, serum; S.D., standard score of values with respect to our reference population (11) expressed in standard deviation units; ANUC, analytic ultracentrifugal values calculated from regression formulae (11) using TC values alone. Thus, ANUC values for HDL = 87 + 0.784x, VLDL = -6 + 0.859x and LDL = 5 + 0.882x, where x values are the calculated agarose concentrations internally standardized by TC alone.

ultracentrifugation. Without the centrifugation some 400-500 serum samples per week could be analyzed. Considering the above, a rough estimate of the unit cost exclusive of the serum cholesterol and triglyceride determinations would be approximately $5.00 - $10.00 per test.

Although we use a specific configuration of hardware, others should work as well or better. Also, with slight modifications, different systems should be able to utilize the substantial software developed for our system. Without going into detailed logistics of large scale screening, the average sample will require 3-4 minutes of processing which is a throughput potential comparable to a typical clinical autoanalyzer.

OTHER SIMPLIFIED QUANTITATIVE LIPOPROTEIN MEASUREMENTS

Alternatives to quantitative electrophoresis include two quantitative lipoprotein procedures based upon lipoprotein precipitation by heparin-$MnCl_2$. The procedure utilized by the Lipid Research Clinics (14) first removes VLDL by preparative ultracentrifugation, then precipitates LDL and calculates all lipoprotein concentrations in terms of cholesterol content of VLDL, LDL and HDL. This procedure does not lend itself to large scale application. A more simplified procedure by Friedewald et al. (15) eliminates the centrifugation step and estimates VLDL cholesterol directly as TG/5. However, samples with TG > 300 cannot be properly analyzed. After both LDL and VLDL are precipitated, LDL and HDL cholesterol are determined. Thus, the Friedewald technique also has the capability for large scale quantitative lipoprotein screening. However, limitations of both precipitation procedures are the necessity of performing accurate cholesterol analyses on dilute lipoprotein containing solutions.

Recently the Friedewald procedure and our automated microelectrophoresis procedure were compared (16) using the analytic ultracentrifuge as a standard (17). Thirty seven normal adult women were studied. Correlation coefficients relating ultracentrifugal and the Friedewald lipoprotein data were 0.849, 0.914 and 0.895 for HDL, VLDL, and LDL, respectively. Analogous correlations for quantitative electrophoresis were 0.929, 0.890 and 0.892, respectively. Table I gives a more detailed comparison of the two methods, including regression formulae for both procedures allowing calculation of equivalent ultracentrifuge data. Comparison of the error in estimating the assumed true value (the analytic ultracentrifuge results) is given by Sy.x. Although agarose provides a somewhat better estimation for HDL, no definitive advantage in accuracy is provided by either method.

However, there are advantages and limitations to both procedures. The Friedewald procedure can be used on frozen sera but requires 3 ml of sera. In contrast, the electrophoresis system must be used with fresh serum but requires only some 100 μl serum if an enzymatic autoanalyzer technique is used for the cholesterol and triglyceride determinations. Thus, such an electrophoresis system would be most useful in screening young children or in studies where only small blood samples can be obtained. Another consideration is that the electrophoretic slides may be quantitatively analyzed several years retrospectively. In balance, it would appear that the electrophoretic method offers advantages for a mass screening program and is comparable in accuracy to an alternate chemical precipitation procedure.

Table I

Comparison of Electrophoretic and $MnCl_2$-Heparin

Procedures with Analytic Ultracentrifugation[a]

Variable	Mean ± SD mg/100 ml	r	$S_{y.x}$	b	a
Agarose HDL	438 ± 111	0.929	33	43	0.717
Agarose VLDL	77 ± 57	0.870	26	2	0.770
Agarose LDL	466 ± 138	0.892	51	27	0.702
$MnCl_2$-Hep. HDL[b]	355 ± 93	0.849	47	78	0.785
$MnCl_2$-Hep. VLDL[c]	158 ± 67	0.914	21	-51	0.686
$MnCl_2$-Hep. LDL[d]	398 ± 106	0.895	50	-21	0.920

[a] $y = b + ax$, where y = analytic ultracentrifuge lipoprotein values and x values are agarose concentrations calculated by both TC and TG internal standardization (5).

[b] $HDL = 5.95 \times HDL_{TC}$.

[c] $VLDL = \text{serum TG}/5 \times 7.30$.

[d] $LDL = (\text{serum TC-HDL}_{TC} - TG/5)\ 3.01$.

Reprinted from reference 16 by permission of the Americal Oil Chemists' Society.

THE FUTURE FOR QUANTITATIVE LIPOPROTEIN ELECTROPHORESIS

Certainly now is not the time to abandon electrophoresis. It is rather a time to increase efforts to improve lipoprotein quantification by this useful simple technique, and to test its potential for improving the definition and treatment of lipoprotein abnormalities. Such improved electrophoretic procedures should enhance diagnostic capability.

The case for intensifying rather than minimizing interest in lipoprotein electrophoresis can be summarized as follows:

If in only 10% of screening cases additional information is provided by quantitative electrophoresis, it is still worthwhile to consider electrophoresis on all cases, because a priori one cannot identify all those 10% of cases that would benefit from electrophoresis. An analogous medical situation is that of remotely suspected bone fractures. A physician would prudently recommend an x-ray in order to properly identify those very few patients who might have bona fide fractures.

Quantitative lipoprotein electrophoresis can provide concentrations of the major lipoprotein classes and these procedures and calculations can be conveniently automated (12).

Complete lipoprotein information is more meaningful and important metabolically than plasma triglyceride and cholesterol information. It seems obvious that these lipoprotein structures that exist naturally in the bloodstream are more important as objects of study than the two major lipid classes that have been extracted from all the lipoproteins (atherogenic and non-atherogenic) and lumped together as a total entity. Lipoproteins are the actively metabolizing units and metabolic defects are reflected in the entire plasma lipoprotein pattern. It is, therefore, logical that the lipoproteins are the important biological units to be studied and quantified.

Last and not the least important is that widespread discontinuance of electrophoresis will have an inhibitory effect of "basic research" which presently is attempting to improve and quantify lipoproteins by electrophoresis. An additional unfavorable effect will be within the practical clinical realm which has been aided by "off the shelf" commercial electrophoresis kits. Since most businesses operate on relatively short-term market expectations, some or most of these kits may be permanently withdrawn from the market if the immediate forecast for electrophoresis is negative.

The blood lipoprotein system is a complicated one and this is emphasized by the present inability to fit all lipoprotein abnormalities neatly into six types. It would appear essential that an improved classification will necessarily involve quantification of all lipoprotein classes. The most inexpensive, convenient, and simple method to do this is by quantitative lipoprotein electrophoresis.

ACKNOWLEDGEMENTS

This work was supported in part by National Institutes of Health grant HL-18574-02 and the U.S. Energy Research and Development Administration. We thank Suzanne Pan for able computer analysis of data, Gerald L. Adamson for excellent technical work and Dorothy S. Sprague for help in preparation of the manuscript.

REFERENCES

1. Beaumont, J. L., Carlson, L. A., Cooper, G. R., Fejbar, Z., Fredrickson, D. S. and Strassen, T. Bull. Wld. Hlth. Org. 43: 891, 1970.
2. Immarino, R. M. Human Path. 5: 626, 1974.
3. Fredrickson, D. S. Circulation 51: 209, 1975.
4. Dyerberg, J. and Hjorne, H. Acta Med. Scand. 191: 413, 1972.
5. Hatch, F. T., Lindgren, F. T., Adamson, G. L., Jensen, L. C., Wong, A. W. and Levy, R. I. J. Lab. Clin. Med. 81: 946, 1973.
6. Miller, G. J. and Miller, N. E. Lancet 1: 16, 1975.
7. Castelli, W. P., Doyle, J. T., Gordon, T., Hames, C., Hulley, S. B., Kagan, A., McGee, D., Vicic, W. J. and Zukel, W. J. Circulation 52 Suppl. II: II-378, 1975.
8. Wood, P. D., Stern, M. P., Silvers, A., Reaven, G. M. and von der Groeben, J. Circulation 45: 114, 1972.
9. Fredrickson, D. S., Morganroth, J. and Levy, R. I. Ann. Intern. Med. 14: 815, 1975.
10. Fredrickson, D. S., Levy, R. I., Bonnell, M. and Ernst, N. DHEW Publication No (NIH) 73-110, 1973.
11. Lindgren, F. T., Adamson, G. L., Jensen, L. C. and Wood, P. D. Lipids 10: 750, 1975.
12. Wong, R. A., Banchero, P. G., Jensen, L. C., Pan, S. S., Adamson, G. L. and Lindgren, F. T. J. Lab. Clin. Med., in press.

13. Lindgren, F. T. In: Analysis of Lipids and Lipoproteins, E. G. Perkins, ed., J. Amer. Oil Chem. Soc., Champaign, IL., 1972, pp. 187-274.

14. Manual of Laboratory Operations, Lipid Research Clinics Program, Vol. 1, Lipid and Lipoprotein Analysis DHEW Publication No (NIH) 75-628, 1975.

15. Friedewald, W. T., Levy, R. I. and Fredrickson, D. S. Clin. Chem. 18: 499, 1972.

16. Lindgren, F. T., Silvers, A., Jutagir, R., Layshot, L. and Bradley, D. D. Lipids 12: 278, 1977.

17. Lindgren, F. T., Jensen, L. C. and Hatch, F. T. In: Blood Lipids and Lipoproteins, G. J. Nelson, ed., John Wiley-Interscience, N.Y., 1972, pp. 181-274.

SOLUTION PROPERTIES OF APOLIPOPROTEINS

Gerhard M. Kostner

Institute of Medical Biochemistry
University of Graz
A-8010 Graz, Harrachgasse 21, Austria

INTRODUCTION

We have learned from Dr. Lindgrens presentation that human serum lipoproteins are prepared customarily by preparative ultracentrifugation. Quantitation of individual density classes can than be performed with the analytical instrument. Thus we are so far only familiar with the operational classification of lipoproteins based on the method used for their separation, reflecting the hydrated densities. As it became evident from numerous publications of the past, and this has been clearly demonstrated at the Protides Meeting of this year, there exist a great variety of additional isolation and subfractionation procedures for serum lipoproteins. In fact we are able now to subfractionate lipoproteins not only according to their flotation behaviour in salt solutions of increased densities, but also according to differences in the protein moiety. Before going into the sometimes troublesome matter of purifying apolipoproteins it proved to be of great advantage to separate first lipoprotein subspecies or families. For that purpose it is necessary to apply not only one or two consecutive separation steps but rather a combination of some 3-6 different ones. After isolation of a maximum number of subfractions differing from each other by one or more apolipoprotein polypeptides they should be delipidated and further processed. This is particularly true since apolipoproteins once present in the lipid free state, tend to aggregate not only each apoprotein with the same species but also with all kinds of polypeptides from other families. Thus aggregates are formed with high molecular weights which in many cases cannot be dissociated quantitatively. Part of this presentation will therefore be devoted to the subfractionation of individual lipoprotein classes on a preparative scale followed by the description of the most effective and convenient delipidation procedures. Only the last part of this chapter deals with the solubility properties of individual apolipoprotein polypeptides - a matter which today is only poorly investigated. Before going into this field let us just summarize the present state of knowledge concerning the identity of apolipoprotein polypeptides of human serum.

TABLE I

Apolipoproteins of Human serum

Apo-lipoprotein	Synonym	Density Class	Mol.wt. x 10^{-3}	References
AI	R-GlnI	HDL	28	(1,2)
AII	R-GlnII	HDL	17	(1,2)
AIII	D	HDL	21	(3,4)
B	Apo-ß	LDL,VLDL	275	(5)
CI	R-Ser	VLDL,HDL	7	(6)
CII	R-Glu	VLDL,HDL	8.5	(6)
$CIII_1$	$R\text{-}Ala_1$	VLDL,HDL	8.5	(6)
$CIII_2$	$R\text{-}Ala_2$	VLDL,HDL	8.5	(6)
D	AIII	HDL	21	(3,4)
E	A.R.P.	VLDL,HDL	39	(7)
a	apoLp(a)	HDL_1	250-300	(8,9)
GRP	---	HDL	?	(10)
D-2	---	HDL	7	(11)

The family classification system proposed by Alaupovic (12) is nowadays widely accepted to denominate lipoproteins and apolipoproteins. Since not all polypeptides of human serum lipoproteins described so far can be integrated into this scheme, trivial names must be used in addition in this report. According to the family concept, apolipoproteins belonging to one family are characterized by their simultaneous occurance into one lipoprotein particle. Triglyceride rich lipoproteins form aggregates of lipoprotein families. Above a hydrated density of approximately 1.025 or after removal of neutral lipids, lipoprotein entities can be prepared which meet more or less closely the requirements demanded for "families". It should however, be emphasized that the in vivo situation or the physiological significance of the described lipoprotein families have never been unequivocally demonstrated so far.

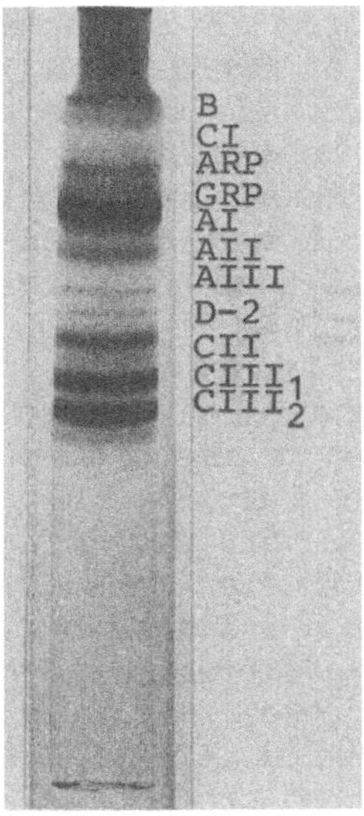

Fig. 1: Disc electrophoresis of 0,1 mg of apoVLDL plus apoHDL in a 10% gel containing 8M urea. Staining was performed with Coomassie blue.

The two peptides listed at the bottom of Tab. 1 have only been partially characterized and will therefore not be considered in this report.

The best characterization of known or even new polypeptides is the determination of amino acid composition and of terminal amino acids. Since this characterization cannot always be carried out, apolipoproteins are most conveniently characterized by their migration rate in defined polyacryl amide gel systems and/or by their immunochemical properties. Fig. 1 shows the polyacryl amide gel electrophoresis pattern of a mixture of apo-VLDL plus apo-HDL. The apoproteins were electrophoresed in 10 % polyacryl amide gels (PAGE) containing 8M urea. The particular gel composition has been described in earlier publications (2,3) and with some experiance there are little difficulties to correlate the stained bands with the individual apolipoproteins. In addition, monospecific antisera which unfortunately are not commercially available proved to be of great importance to characterize and also quantitate polypeptides (13). This is particularly true for apoB, the major constituent of LDL, which does not migrate into the separating gels using the system described for Fig. 1. Since the immunochemical behaviour of intact lipoproteins and their lipid free counterparts may vary considerably, all immunochemical characterizations should be performed with both materials.

Isolation and Subfractionation of Serum Lipoproteins

Although it is not the aim of this presentation to introduce the preparation of lipoprotein subfractions, this matter will still be overviewed in order to be able to follow the subsequent explanations. Intact lipoproteins should be fractionated as far as possible and the separation of apolipoproteins should proceed with the obtained material. The first step for preparing large amounts of apolipoproteins most commonly are precipitation methods. Although a great variety of substances have been described for more or less selective precipitation of lipoproteins, only few of them can safely be applied without alterations of the obtained material. From all the precipitating agents described by Burstein (14) I feel that sodium phosphotungstate can be recommended mostly for precipitation individually VLDL, LDL and HDL. After solubilisation of the precipitate in 0.1M sodium citrate buffer, lipoproteins are further purified by preparative ultracentrifugation. Another very promising procedure, recently introduced by Garcia (15) fractionates individual lipoprotein classes by ammonium sulfate precipitation followed by sodium phosphotungstate. In this way the preparative ultracentrifuge can be completely avoided. Both methods yield lipoprotein mixtures which are essentially free of nonlipoprotein components. A further subfractionation of these is possible by many selected procedures. The most effective ones are immunoadsorbers, lectines linked to a solid matrix e.g. Sepharose 4B, and preparative column chromatography using Biogels or hydroxyapatite. By these methods, we and others were able to subfractionate intact lipoproteins into numerous distinct entities differing from each others not only by the lipid but also by the protein moiety. Some of the subclasses represent lipoprotein families characterized by the presence if apolipoproteins belonging to only LpA, LpB or LpC. As an example I briefly will describe the method which has been successfully applied in our laboratory for the isolation of the three major lipoprotein families.

Isolation of lipoproteinB (LpB). Although LpB represents more than 95 % of the lipoprotein mass of LDL, it is sometimes very cumbersome to prepare LpB uncontaminated by apoA and apoC polypeptides. Yet a highly purified product is absolutely necessary for studying the structure of apoB. In our procedure we start with pooled serum from fasting normals. Immediately after sampling, preservatives and protease inhibitors (Na_2EDTA, NaN_3, Merthiolate, Chloramphenicol and phenyl methyl sulfonyl fluoride) should be added. In order to avoid troubles expressed by the formation of aggregates it is of advantage to start with serum from Lp(a) negative donors. After centrifugation in a laboratory centrifuge VLDL and LDL are precipitated out from the serum according to Burstein and Morfin (16). The precipitate is removed at 8000 g, washed once with saline containing sodium phosphotungstate and $MgCl_2$ of the same ratio as used in the first step, and suspended in 1% sodium citrate. This suspension is dialysed overnight against saline preferentially at 4^oC. All solutions used throughout the isolation procedure should contain some of the preservatives listed above. The density of the dialysed solution is adjusted to 1.025 by the addition of solid NaCl and VLDL and LDL_1 are floated in the preparative ultracentrifuge at 120 000 g for 18 hrs and 15°C. The sedimenting material is removed by tube slicing and adjusted to a density of 1.055 with NaCl. LDL are floated at 125 000 g for 22 hrs and 15°C. The yellow layers on top of each tube are mixed, diluted with Tris.HCl buffered saline (pH 8.5) to a final LDL concentration of 10 mg/ml and chromatographed in portions of 8-10 ml over a column, 100 x 2.5 cm, packed with Biogel A-5m in Tris-HCl buffered saline. The elution pattern of a representative experiment is shown in Fig. 2. The main peak contains LpB with a purity exceeding 98.5 %, while peak 1 are aggregates and peak 3 LpA and LpC. The use of Tris.HCl buffer seems to be essential for removal of most of the apoA and C peptides since other substances as for example Barbital-Na or $NaHCO_3$ were unable in our hands to dissociate the LpB - LpA - LpC - complex.

Isolation of LpA and LpC from HDL. LpA is the major lipoprotein family of HDL. This density class is therefore used for the LpA preparation. In most of the cases we start with HDL_3 as this material represent more than two third of the total HDL. In addition HDL_2 does not always yield pure LpA. HDL is prepared from the supernatant of the VLDL-LDL precipitation step by ultracentrifugation at d = 1.22, adjusted with solid NaBr, for 24 hrs

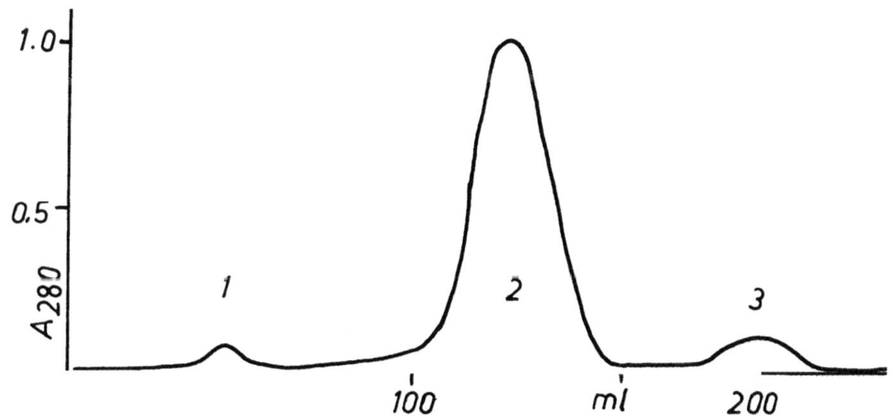

Fig.2: Elution pattern of low density lipoproteins (d 1.025 - 1.055) from a column, 2.5 x 100 cm packed with Biogel A-5m in 0.15M NaCl containing 0.05M/l Tris.HCl, pH 8.5. 100 mg LDL were applied.

at 145 000 g and 15°C. If larger volumes have to be processed, HDL are precipitated also with sodium phosphotungstate (16). Either HDL preparation is dialysed against a NaBr solution of d = 1.110 and ultracentrifuged at 145 000 g for 30 hrs. The bottom fraction (two third of the tubes) represent HDL_3 while HDL_2 floate under these conditions. Since HDL_3 is the major substrate for the enzyme lecithin:cholesterol acyl transferase (LCAT) we add to the starting serum the inactivator Ellman's reagent (10mM/l). HDL_3 is dialysed against 0.05 M potassium phosphate buffer (KPP), pH 6.8, and passed over a column packed with hydroxyapatite in the same buffer. Hydroxyapatite prepared according to the procedure described at the Protides meeting (17) should be used in order to obtain good flow rates. LpA, representing more than 60% of the applied material, elutes with the 0.05 M KPP. With 0.1 - 0.2M KPP some aggregation products of LpA with LpC are eluted while a lipoprotein with only apoC peptides is obtained with 0.25 - 0.35 M KPP. The so obtained lipoprotein families are used as starting material for preparation of apolipoproteins.

Preparation of Apolipoproteins

Since the solubility properties of apolipoproteins are reflected most drastically by the delipidation procedure applied, this matter will be covered here in more detail. An enormous variety of delipidation methods have been described in the literature and there are probably only few of them which have not been tried in my laboratory. It tourned out that only the minority of them can be recommended for general use. I will restrict my presentation to those only which I feel are the most convenient and effective ones. A more detailed review of this subject can be found in Ref. (18). In general, two different possibilities exist to obtain apolipoproteins: Delipidation of lipoproteins in solution and delipidation of freeze dryed material.

Delipidation of lipoproteins in solution. These methods yield apolipoproteins with very good solubility properties and are particularly recommended for preparation of apoB. A very safe method is that described by Shore and Shore (19) using diethylether and ethanol -diethylether at 0-4°C. Lipoproteins are dialysed against 0.5 M NaCl, 0.01 M Tris, pH 7.6, 0.0008 M EDTA and extracted first with diethyl ether and than with ethanol: diethyl ether 2:1, until a lipid free apolipoprotein suspension is obtained. The major drawbacks of this method are the solubility of the small peptides especially of the C- family in the organic phase and the accumulation of large volumes if greater amounts of lipoproteins are delipidated. Another possibility for delipidation is the use of detergents. Helenius and Simons (20) compared the effectivity of Nonident P40, Triton x 100, sodium dodecyl sulfate (SDS) and

Na-desoxycholate (NaDOC) for the preparation of apoB. The best results were obtained with NaDOC. 11.5 mg of this substance are added to each mg of LDL at pH 10, adjusted with Na_2CO_3, and the mixture is chromatographed over Sephadex G200 or Biogel. The first peak represents apo-LDL while lipid micelles are eluted in a later fraction.

Another very efficient method is that described by Kane (21) using tetramethyl urea (TMU), especially if small amounts of lipoprotein solutions are to be delipidated. Although this method delipidates quantitatively only VLDL and LDL, it can be used also for HDL if samples are investigated in PAGE in the presence of 8M urea. The lipoprotein solutions are mixed with equal volumes of TMU and immediately electrophoresced in 8M urea containing gels. ApoAI may form aggregates if higher concentrations are present, which eventually interfere with the peptide band pattern. This method has also been successfully applied to quantitate the apoB content of VLDL or LDL in metabolic investigations. ApoB namely precipitates out quantitatively under the described conditions while ApoA and apoC remain in solutions, thogether with the lipid part. Though it is not possible to separate all peptides from the lipid moiety with TMU.

To circumvent these problems, we searched for delipidating agents which are non-miscible with aqueous buffers and separate lipids from proteins in an organic phase. One of the most powerful effective and milde substance with that respect was found to be glycerol-1-hexyl ether (GHE). The synthesis of GHE has been described in detail at the last year EFRAC meeting in Brugge (22): Isopropylidene glycerol and hexyl bromide both in benzene are mixed at a molar ratio of 2:1 and boiled under reflux in a water separator. Solid KOH is added after one hr and the whole mixture is poored on crashed ice. The formed GHE is extracted with diethyl ether and distilled under reduced pressure (BP at 0.2 mm Hg: 88°C). GHE is a colorless, harmless liquid which can be stored at room temperature for years without loss of its property. Upon the addition to lipoprotein solutions, GHE removes immediately all lipids from VLDL and LDL. HDL are only partially delipidated. Pretreatment of HDL with phospholipase-C yields also totally delipidated apoHDL.

Delipidation with GHE. For delipidation of VLDL, LDL or LpB, 0.1 - 10 ml of solution containing 1-30 mg of lipoproteins per ml of 0.15 M NaCl are mixed at room temperature with 10 % of the volume with GHE and shaken for 30 min. vigorously with a Vortex mixer. The mixture is centrifuged for 10 min. at 5000 g and the organic phase at the top is aspirated by a syringe. Addition of diethyl ether at this step may save 50 % of the GHE. The extraction procedure is repeated once or twice. The formed apoB accumulates at the interphase while apoA and apoC peptides remain mostly in solution. HDL are preincubated at an enzyme : substrate ratio of 1:2000 for

12 hrs at 37°C in the presence of 0.05 M Ca^{++} and delipidated as described for LpB. The described method handles any amount of lipoprotein and works extremely rapid. It is recommended primarily for delipidation of LpB and for serial analyses of any lipoprotein fraction in small volumes.

In addition to the methods described here for delipidation of lipoproteins in solutions, a great deal of other procedures have been described in the literature. Some authors work with isobutanol: acetic acid: water mixtures and chromatograph this mixture over Sephadex LH 20, others are using chloroform methanol. We however consider the GHE, NaDOC and TMU methods as the most effective ones.

Delipidation of freeze dryed lipoproteins. For some purpose the delipidation of lyophilized lipoprotein fractions should be prefered to the methods described in the former paragraph. The major advantages are that all apolipoproteins definitely remain insoluble in the organic phase. In addition the accumulation of large volumes can be avoided if one works with large amounts of material. It should, however be noted that apoB in these cases is very hard to solubilize even in buffers containing detergents. We have successfully applied the following method for delipidation of chylomicrons, VLDL and HDL: Lipoprotein fractions are dialysed eshaustively against a solution of 0.001 M $NaHCO_3$ containing 0.01 mg/ml of Na_2EDTA. The volume and concentration of the solution is not crucial. After equilibration with the buffer in the cold room the solution is lyophilized. Triglyceride rich lipoproteins but also huge amounts of HDL are extracted first with peroxide free diethyl ether at room temperature. This step eventually is performed in a Soxhlet apparatus. Alternatively centrifuge tubes with glass stoppers are used. The extraction with diethyl ether is repeated several times until an aliquot of the extract gives no residual lipid after evaporation. In the next step ethanol: diethyl ether 3:1 is used. This extraction step has to be carried out in the cold room and care has to be taken that by no means the temperature of the lipoprotein raises over 4°C. The amount of solvent and number of extraction steps must be adapted individually. HDL or LpA has to be extracted 5-6 times, VLDL and chylomicrons 3-4 times. After the removal of all polar lipids two to three more extractions with diethyl ether follow. The obtained white material is dryed under nitrogen in a desiccator. Water free apolipoproteins can be stored at room temperature or 4°C.

Antigenicity of Delipidated Apolipoproteins

For the location of apolipoproteins during their separation but also for their determination in individual lipoprotein subfractions immunochemical tests are widely used in addition to other analytical methods as for example PAGE. There seems to be little doubt that the immunochemical behaviour of most if not of all polypeptides differ after delipidation from that in intact lipoproteins. This phenomenon has clearly been demonstrated for example with apoAI, which is displaced in the radio-immunoassay by the corresponding amount of LpA only to 5 % (23). In addition it has been shown that the C- and N-terminal region of that apolipoprotein has different antigenic properties. For quantitative immunochemical tests therefore, variable results can be obtained with antisera from different species. For qualitative tests all apolipoproteins are solubilized best in 0.05 M Tris.HCl, pH 8.0, containing 8 M urea and investigated by immunoelectrophoresis or immunodiffusion. The shape of the precipitin arc in immunoelectrophoresis using a specific antiserum may vary from intact to delipidated lipoproteins. In immuno diffusion experiments however for all apolipoproteins a reaction of complete identity has been obtained in our laboratory. ApoB solely, gives only immunochemical reactions in the presence of certain detergents as for example SDS or NaDOC. With Triton X100 on the other hand-though capable to solubilize apoB - no precipitation using anti-LpB has been observed. Applying SDS another unspecific precipitate in addition to the specific precipitine line is formed which is likely to mislaed unexperienced researchers.

Solution Properties Of Apoproteins

As pointed out above, the properties of apoproteins depend to a major degree upon the isolation and delipidation procedures. Generalizing matters on can say that all polypeptides investigated so far get more and more insoluble if repeated freeze drying is performed or if extreme pH values are applied. In addition it has been found that mixtures of nonidentical apolipoproteins in most cases are much better soluble than purified polypeptides. The exact condiiton for the most effective solubilization of all apolipoproteins individually is only poorly investigated. The statements made in the following are therefore restricted to some limited investigations carried out in our laboratory or partly also published by other research groups. The solubility properties of apolipoproteins can be drastically altered on chemical modification. The most frequently used methods are acetylation, maleylation, succinylation and citraconylation. The latter has the particular advantage, that the covalently linked ligands can be removed quantitatively in a short period of time at pH 4.0 or bellow. The following report, however, deals only with untreated apoprotein samples. The solubility properties in addition are influenced markedly by the presence of residual lipids-especially by phospholipids. This seems to be the main reason why divergent reports have been published from several laboratories. In our studies we have checked the purity of apolipoproteins for residual neutral lipids and phospholipids. The phospholipid content in all cases was less than 0.4 % by weight while other lipids were absent.

Solubility of apoB. Apolipoprotein-B requires the strongest detergents for solubilisation. This seems to be the reason why the exact number and molecular weight of apoB polypeptide is still a matter of speculation. What makes matters worse is the fact that apoB is prone to enzymatic degradation by proteases of endogenous or bacterial origins. The degraded material has other solubility properties as the intact B-protein.

ApoB is very sensitive to the delipidation methods. After delipidation in solution applying the methods described above, apoB gives clear solutions in organic acids such as 50 % acetic acid, 80 % formic acid or others. Centrifugation of these products at 500 000 g min., however, yields pellets of aggregated apoB. The supernatant contain small amounts of possibly "monomeric" apoB. In the presence of large amounts of A or C-peptides, apoB can even be solubilised in 8M urea or 6M guanidinium hydrochloride (Gu.HCl). Much better solubilisation of apoB is achieved with SDS, sodium decyl sulfate (SDeS) or NaDOC. If the latter detergent is used the delipidation method of Helenius and Simons has to be applied (20). With SDS highly concentrated apoB solutions exceeding 20 mg/ml can be obtained. The presence of reducing agents e.g. mercapto ethanol or DTT is not obligatorybut helps to avoid aggregation. To 1 mg of apoB 8 mg of SDS must be added in addition to the desired amount of buffers. The mixture is heated for 3 min at 100°C for inactivation of proteases and stirred under nitrogen until a clear solution is formed. Dialysis can now been applied against buffers containing

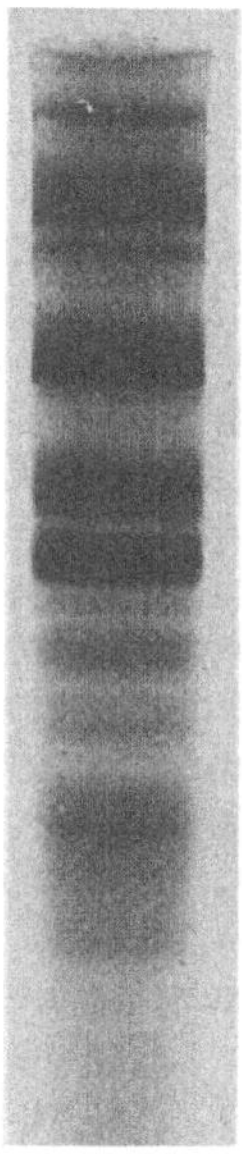

Fig. 3:

Polyacryl amide gel electrophoresis of apolipoprotein-B in a 5 % gel containing 0.1 % SDS. Staining was performed with Coomassie blue.

0.1 % SDS or 0.01 % SDeS. Further removal of detergents by dialysis leads to the precipitation of apoB. Fig. 3 shows the PAGE pattern of such apoB in 5 % gels, 0.1 % SDS. Although all mentioned preservatives were present during preparation it cannot be excluded that we are dealing here with a secondary pattern created by proteases.

Solubility properties of apoA peptides. The major work concerning LpA has been performed on apoAI and AII and no systematic analyses on apoAIII are known. ApoAI and most probably also AIII are single stranded polypeptides while AII is composed of two identical subunits linked with disulfide bonds. ApoAI and AII have been found to undergo selfaggregation in all kinds of buffers and solvents. From their role lipid binding proteins one would expect that these interactions are caused by hydrophobic forces. This assumption is underlined by the findings that anionic and cationic detergents bind to apoAI and AII with the same number of sites (i.e. 3-4) and with the same order of magnitude (association constants $2x10^4$ litres/mol) (24), and further that saturation of the hydrophobic regions counteract the aggregate formations.
ApoAI and AII exhibit marked secondary and tertiary structures in ordinary buffers, which are temperature dependent and easy to destroy not only by detergents but also at exteme pH values i.e. bellow 2 and above 12, (25), or substances such as Gu.HCl or urea. From all these observations it is easy to understand that the solubility properties of apoAI and AII are mostly reflected by their formed structure which is subject to change under various experimental conditions. This process is not reversible under all circumstances. In Tris-HCl buffer, pH 8.3, ionic strenght 0.05, apoAI forms dimers and apoAII tetramers. Addition of Gu.HCl or other dissociating agents to apoAI leads to formation of the monomeric peptide. ApoAII in addition requires the presence of mercapto ethanol or DTT. Summarising the results one can say that apoAI and AII are well soluble in slightly alkaline buffers though tend to form aggregates of dimers or oligomers. By the addition of detergents or dissociating agents such as SDS, SDeS, Gu-HCl or urea, monomers are formed.

Organic acids have also successfully applied for the solubilisation of LpA peptides. 1-2 M acetic acid is a very effective solvent and was used for isolation of individual LpA polypeptides. At concentrations of 10 mg/ml or less most of the A- and C-peptides seem to be present in monomeric forms and can be conveniently separated over Sephadex G75 - G200 or other molecular sieves (2,3). Immediately after separation the fractions can be lyophilized. It should be noted here that only analytical grade reagents should be used throughout all investigations of apolipoproteins.
For apoAIII (apoD) essentially the same statements can be made as for AI and AII though exact investigations on this protein are missing so far.

Solubility properties of apoC peptides. All apoC polypeptides have a much lower molecular weight comparing to apoA. Without doubt apoC peptides do have also hydrophobic regions in their primary structure which have clearly demonstrated by Gotto's group (26). Detergents have therefore frequently been used for their purification. The hydrophobicity of C peptide might also be reflected by their solubility in organic solvents e.g. ethanol - diethyl ether - water mixtures. Profound investigations on the solubility of apoC peptides are not known so far. This report therefore is restricted to own observations. For apoC we made the same observations as for apoA namely that with the progress of purification their solubility gets worse. This is especially so in buffers containing no dissociating agents. Nevertheless in most cases monomeric forms seem to exist in buffers containing 6-8M urea of pH 7-9. Repeated freeze drying and other processing renders apoCI, CII and CIII in a form which tends to give aggregates in ordinary buffers. ApoC is excellently soluble in 1-2M acetic acid. Extreme pH values and detergents certainly change the secondary and tertiary structure of apoC in the same manner as observed for apoA. After removal of these substances aggregate formations are more likely to occur as compared to untreated apoC.

Solubility properties of other apolipoproteins. One major constituent comprising approximately 15% of the VLDL protein is the "arginin rich" polypeptide (A.R.P.) also called apoE. Nevertheless this peptide has been characterized only very recently (7). This most probably was caused by the fact that A.R.P. is almost insoluble in the absence of dissociating agents or detergents. Monomeric A.R.P. solutions may only be formed in the presence of 6M Gu.HCl, SDS, or possibly also 8M urea. We have repeatedly used 1 M acetic acid for isolation of A.R.P. over Sephadex G100. The latter two chemicals however seem to be to weak to be used for solubilization of isolated lyophilized A.R.P.
Another apolipoprotein which should be considered here is the "a" peptide the characteristic antigen of Lp(a). It differs widely from all proteins mentioned so far since it is a glycoprotein with more than 30 % carbohydrates by weight. The major portion of carbohydrates is neuraminic acid. The exact molecular weight of the "a" peptide is not known but it seem to exceed 250 000 daltons (9). Because of the high neuraminic acid content, the a-protein does not aggregate in ordinary buffers and neither urea nor detergents are necessary for its solubilisation. After delipidation of Lp(a) the antigen can be extracted with 0.1M Tris.HCl, pH 8.2. A certain portion however, depending on the isolation procedure, remains complexed with apoB. This part cannot be extracted even not with the strongest dissociating agents. The pattern of apo-a in PAGE in the presence of 8M urea is essentially the same as without urea. In SDS containing gels the same kind of pattern is observed as compared to the former gel system. Much work remains to be done in order understand the

kind of interactions of apo-a with LpB giving rise to the formation of the Lp(a) lipoprotein.

SUMMARY AND CONCLUSION

Human serum lipoproteins are constituted of a minimum of then nonidentical apolipoproteins. Alignement of these peptides with increasing solubility in ordinary buffers should give the following picture: ApoB $\ll$ A.R.P. $<$ apoAIII $<$ apoAI $<$ apoAII $=$ apoCII $<$ apoCI $<$ apoCIII $\ll$ apo-a. Excepting apo-a the same figure is obtained if peptides are listed with decreasing molecular weight. It has been found that the solution properties of almost all apoproteins is mostly reflected by the procedures applied for their purification and even more by the delipidation method. For the polypeptides apoB, A.R.P. and possibly also apoAIII delipidation of lipoprotein solutions using TMU, GHE or detergents are recommended to yield fractions with optimal solution properties. Other peptides may be delipidated with diethyl ether and ethanol out of solutions or of freeze dryed lipoproteins. The quality of the delipidated material in many cases can be easily checked by inspection of the pattern in PAGE or the precipitin arcs in immunoelectrophoresis. The occurance of single sharp bands without major staining of gel portions between bands in PAGE suggest very good solubility. The observation of smeary bands on the other hand indicates partial alterations of the material. Generalizing matters on may say that apolipoproteins obtained directly after mild delipidation are very well soluble in most cases even in the absence of dissiciating agents (excepting apo-B). As the purification of distinct polypeptides proceeds the obtained material tends to aggregate even in the presence of urea or acetic acid. In ordinary buffers all the apolipoproteins listed above excepting apo-a tend to aggregate, forming dimers and oligomers. The forces causing this associations are of hydrophobic nature but also hydrogen bonds are likely to occur. Ionic detergents especially SDS or SDeS are capable to render the aggregates in monomeric forms. Since the latter are not easy to remove quantitatively, other substances are frequently used. The strongest one in that sense seems to be Gu.HCl which gives in all cases monomeric solutions excluding apoB. Less powerful but still very frequently applied are 8M urea and 1-2M acetic acid. The presence of residual lipid material, especially of lecithin, markedly influences the behaviour of apolipoproteins.

The immunochemical properties of all apolipoproteins present onto intact lipoproteins differs from that in the lipid free state. ApoAI for example has twenty times more antigenic sites than LpAI. Similar observations have been made with other peptides. Nevertheless all apoproteins with the exception of apoB do form immunochemical reactions of identity before and after delipidation. ApoB has to be solubilised in NaDOC or SDS in order to obtain an immunochemical reaction. Nonionic detergents, though capable to solubilise the B protein, are useless for that purposes. Although a great deal of progress has been reached in recent years much more work seems to be necessary to uncover the exact interactions of apolipoproteins with each other and also with different classes of lipids. Some of the methodology outlined in this report may help to do so.

Part of these investigations have been supported by the Österreichischen Forschungsrat and the Österreichischen Nationalbank.

REFERENCES

1) B.Shore and V.Shore, Heterogeneity in protein subunits of human serum high density lipoproteins, Biochemistry 7, 2773-2777 (1968)

2) G.Kostner and P. Alaupovic, Studies of the composition of plasma lipoproteins. C-and N-terminal amino acids of the two non-identical polypeptides of human plasma apolipoprotein A, FEBS Lett. 15, 320-324 (1971)

3) G.Kostner, Studies of the composition and structure of human serum lipoproteins. Isolation and partial characterization of apolipoproteinAIII, Biochim.Biophys.Acta, 336, 383-395 (1974)

4) W.J.McConnathy and P.Alaupovic, Studies on the isolation and partial characterization of apolipoprotein D and lipoprotein D of human plasma, Biochemistry 15, 515-520 (1976)

5) R.Smith, J.R. Dawson and C.Tanford, The size and number of polypeptide chains in human serum low density lipoproteins, J.Biol. Chem. 247, 3376-3381 (1972)

6) W.V.Brown, R.I.Levy and D.S.Fredrickson, Further characterization of apolipoproteins from the human plasma very low density lipoproteins, J.Biol.Chem. 245, 6588-6594 (1970)

7) B.Shore and V.G.Shore, An apolipoprotein preferentially enriched in cholesteryl ester rich very low density lipoproteins, Biochem. Biophys.Res.Commun. 58, 1-7 (1974)

8) G.Utermann and H.Wiegandt, Darstellung und Characterisierung eines Lipoproteins mit Antigenwirksamkeit im Lp-System. Humangenetik 8, 39-46 (1968)

9) G.Jürgens and G.M.Kostner, Studies on the structure of the Lp(a) lipoprotein. Isolation and partial characterization of the Lp(a) specific antigen, Immunogenetics 1, 560-574 (1975)

10) S.O.Olofsson and A.Gustafson, Studies on human serum HDL:II.Isolation of subfractions in the cold, Scand.J.clin.Lab.Invest. 34, 257-265 (1974)

11) C.T.Lim, J.Chung, H.J.Kayden and A.M.Scanu, Apoproteins of

human serum high density lipoproteins. Isolation and characterization of the peptides of Sephadex fraction V from normal subjects and patients with abetalipoproteinemia, Biochim.Biophys Acta 420, 332-341 (1976)

12) P.Alaupovic, Apolipoproteins and lipoproteins, Atherosclerosis 13, 141-146 (1971)

13) G.M.Kostner, J.R.Patsch, S.Sailer, H.Braunsteiner and A.Holasek, Polypeptide distribution of the main lipoprotein density classes separated from human plasma by rate zonal ultracentrifugation, Eur.J.Biochem. 45, 611-621 (1974)

14) M.Burstein, H.R.Scholnik and R.Morfin, Rapid method for the isolation of lipoproteins from human serum by precipitation with polyanions, J.Lipid Res. 11, 583-595 (1970)

15) M.Garcia, Darstellung des gesamten Spektrums der Humanplasmalipoproteine durch Präzipitation, Thesis, University of Marburg/Lahn (1976)

16) M.Burstein and R.Morfin, Precipitation des Alpha Lipoproteins du serum par le Phosphotungstate de Sodium en presence du Chlorure de Magnesium, Life Sci. 8, Part II, 345-348 (1969)

17) G.M.Kostner, The distribution of apolipoproteins in subfractions of normal human serum lipoproteins obtained by various techniques, Prot.Biol.Fluids, XXV Coll. (1977)

18) J.H.Ledford, Thesis: Studies on the delipidation of human serum lipoproteins, Univ.of Oklahoma (1969)

19) V.G.Shore and B.Shore, Heterogeneity of human plasma very low density lipoproteins. Separations of species differing in protein components. Biochemistry 12, 502-507 (1973)

20) A.Helenius and S.W.Simons, Removal of lipids from human plasma low density lipoproteins by detergents, Biochemistry 10, 2542-2547 (1971)

21) J.P.Kane, A rapid electrophoretic technique for identification of subunit species of apoproteins in serum lipoproteins, Analyt.Biochem. 53, 350-364 (1973)

22) G.M.Kostner, Studies on the structure of lipoprotein-B: Preparation of a soluble apolipoprotein-B. In: Lipid-Protein Complexes,Proc. of the EFRAC Conference in Bruges 1966, Plenum press.

23) G.Schonfeldt, R.A.Bradshaw and J.S.Chen, Structure of high density lipoprotein. The immunologic reactivities of the COOH- and NH_2 terminal regions of apolipoproteins AI. J.Biol.Chem. 251,3921-3926 (1976)

24) W.L.Stone and J.A.Reynolds, The Self-association of the apo-GlnI and apo-Gln-II polypeptide of human high density serum

lipoproteins, J.Biol.Chem. 250, 8045-8048 (1975)

25) J.Gwynne, B.Rewer and H.Edelhoch, The molecular properties of apoAI from human high density lipoprotein, J.Biol.Chem. 249, 2411-2416 (1974)

26) J.P.Segrest, R.L.Jackson, J.D.Morisett and A.M.Gotto,Jr. A molecular theory of lipid-protein interactions in the plasma lipoproteins, FEBS Letters 38, 247-253 (1974)

APOLIPOPROTEIN QUANTITATION

Peter N. Herbert,* Linda L. Bausserman,
Lloyd O. Henderson,* Robert J. Heinen,
Marguerite J. LaPiana, Eve C. Church, and
Richard S. Shulman

Molecular Disease Branch, National Heart, Lung, and
Blood Institute, National Institutes of Health,
Bethesda, Maryland
*Division of Clinical and Experimental Atherosclerosis,
The Miriam Hospital, Providence, Rhode Island

The problems encountered in attempts to quantitate the apolipoproteins are common to all efforts to quantify minor constituents of complex mixtures. Considerations of selective and non-selective losses in the preparation of test samples, specificity, sensitivity, precision, and accuracy are applicable to the selection and evaluation of available techniques. The natural association of apolipoproteins with lipids presents additional problems that must be resolved in the validation of a proposed procedure.

Two general approaches to apolipoprotein quantitation will be discussed in this chapter. Estimations based on column chromatographic fractionation and polyacrylamide gel electrophoresis will be considered first. These methods require isolation of the lipoproteins before apoprotein quantitation. Immunological methods including single diffusion, electroimmunoassay and radioimmunoassay will be emphasized because of their applicability to quantitation in both whole plasma and lipoprotein subfractions.

TECHNIQUES APPLICABLE TO ISOLATED LIPOPROTEINS

Preparative Ultracentrifugation. Ideal techniques for lipoprotein isolation are not available. Isopycnic ultracentrifugation in fixed angle rotors is most widely employed (1). Very low density lipoproteins (VLDL) can be recovered by a single centrifugation step at plasma density, but isolation of low density lipoproteins (LDL) and high density lipoproteins (HDL) require a minimum of two ultracentrifugation steps at high salt concentrations. All lipoprotein classes so isolated are usually contaminated with other plasma proteins that may be removed by one or more recentrifugations. Such ultracentrifugal "washing," however, is attended by selective losses of both lipids and proteins and alteration of lipoprotein physical properties (2). Such washing is most useful when apoprotein purification is the goal. Removal of non-apolipoprotein contaminants enhances the likelihood of obtaining individual apoproteins in homogeneous forms. When simple apoprotein quantitation is the goal, however, ultracentrifugal washing is best avoided.

Delipidation of Lipoproteins. The methods for apoprotein quantitation discussed in this section require delipidation prior to application of the quantitative test. This is usually achieved by lipid extraction with organic solvents (3, 4), detergents (5), or tetramethylurea (TMU) (6). When column chromatography has been employed for apoprotein semiquantitation, a variety of methods of organic solvent extraction have been advocated. The most widely used is that recommended by Scanu and Edelstein in which a solvent system of ethanol and diethylether is employed (3). In the authors' experience, the most reliable combination of organic solvents is that of methanol and diethylether. Lipoproteins are dialyzed against 0.15M NaCl, 1 mM EDTA, 5 mM NH_4HCO_3. Two ml or less of the lipoprotein solution are added dropwise to 15 ml of vortexing methanol at 0°C in a 45-50 ml conical tube. The tube is filled immediately with diethylether, inverted several times and placed for 10 minutes in wet ice. The protein is sedimented by low speed centrifugation (e.g., 2,000 rpm in any clinical centrifuge with swinging bucket rotor) for two minutes. The organic solvent is removed by aspiration or decantation, the protein resuspended in methanol, and the tube again filled with diethylether and placed for 10 minutes in wet ice. The protein is pelleted by centrifugation (this time 3,000 rpm for four minutes) and the organic solvent again discarded. Finally, the protein is rinsed twice with diethylether and dried in a thin film on the tube surface with a stream of nitrogen. Residual traces of ether are removed by vacuum suction before resolubilization.

The most frequent reasons for inadequate delipidation, nonquantitative recovery of protein, or recovery of protein with poor solubility characteristics include: (1) failure to dialyze excess KBr from the lipoprotein solution; (2) dialysis of the lipoprotein solution against distilled water rather than 0.15M NaCl; (3) failure to keep organic solvents at 0°C or less; (4) use of prolonged or high speed centrifugation to initially sediment the protein when the delipidation mixture contains the original volume of water (the conditions of centrifugation are much less important after the initial protein sedimentation); and (5) drying the protein in a pellet rather than a thin film. Methanol is the critical component of this organic solvent mixture. Substitution of chloroform for diethylether as the apolar solvent does not improve the degree of delipidation (usually less than 1% residual phospholipid) and renders the protein much less soluble.

Apoprotein Solubilization. Techniques of apoprotein solubilization are discussed in detail elsewhere in this monograph and will be treated only briefly here. ApoVLDL and apoLDL, when totally delipidated and in dry form, require detergents for total resolubilization. Sodium decylsulfate is optimal at a concentration of 0.1-0.2M because it is more easily removed by dialysis. Sodium dodecylsulfate is not as readily dialyzed, but is satisfactory. (Both will be referred to here as SDS). Detergent solubilized apoVLDL and apoLDL can be immediately fractionated by column chromatography in buffers containing urea or low concentrations of SDS (e.g., 2 mM). If chromatography in guanidine HCL-containing buffers is desired, the anionic detergents must first be removed by dialysis because high molarity guanidine HCl will precipitate SDS and some protein.

ApoHDL is readily solubilized in aqueous solutions containing 6-8M urea (7), 4-6M guanidine HCl, 50% acetic acid (8) or 0.1M NH_4OH. Solubilization in simple Tris buffers at pH 8.0 has also been frequently used, but resolubilization in such buffers in sometimes slow and incomplete.

Purified apolipoproteins invariably are more difficult to solubilize than the same apoproteins present in mixtures. Storage in solutions containing urea should be avoided because of the inevitable carbamylation (9, 10), and guanidine HCl buffers are of too high ionic strength to permit gel electrophoresis. We have found the following solutions suitable for completely resolubilizing purified apolipoproteins in concentrated solution: 0.1M NH_4OH for apoproteins A-I, C-II and E; 0.1M NH_4HCO_3 for A-II, C-III-1 and C-III-2; and 1.0M acetic acid for C-I. When apoB from VLDL or LDL is required in concentrated solution, the dilute protein solutions containing 2 mM SDS, 0.1M NH_4HCO_3 are lyophilized, resolubilized in a small volume of 0.1M NH_4OH and excess SDS removed by dialysis. The routine use of volatile buffers permits repeated lyophilization for storage without great increases in salt concentration.

Quantitation by Column Chromatography

ApoVLDL. Gel chromatography has been employed in many laboratories to obtain estimates of the quantities of major protein species in VLDL. No chromatographic medium or buffer system permits resolution of all of the VLDL apoproteins and the choice of chromatographic conditions is dictated in part by the fractions whose quantitation is most relevant to the experimental protocol. ApoVLDL can be resolved into two major and one minor peak on columns of Sephadex G-150 (11) or G-200 (10) (Fig. 1) in buffers containing SDS. The void volume peak (fraction 1, Fig. 1) contains mostly apoB, and the C-proteins comprise the second major peak (fraction 3, Fig. 1). The B protein is relatively pure, but the C-protein fraction is typically contaminated with proteins from the minor peak (fraction 2, Fig. 1) which contains apoE and a number of other apoproteins and contaminating plasma proteins (10).

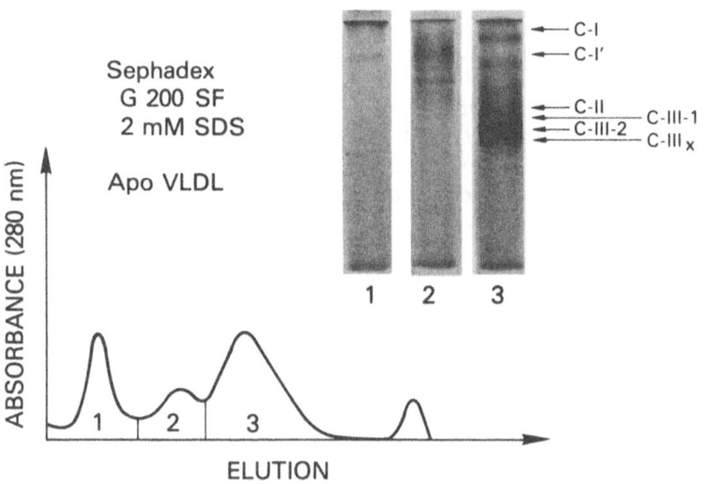

Fig. 1. Sephadex G-200 S.F. chromatography of 20 mg of apoVLDL. A 2.5 x 150 cm column was packed with the glass bead technique and eluted in descending fashion (10). Polyacrylamide gel electrophoresis of the indicated pools was performed in 7.5% acrylamide monomer in the system of Reisfeld and Small (12) modified so that all solutions contained 8M urea. S.M. refers to the starting material and Arabic numbers to the corresponding chromatographic fractions.

Chromatography of apoVLDL on columns of Bio-Gel A 0.5M in 0.1M NH_4HCO_3, 2 mM SDS, pH 7.9, permits even better resolution of the apoB peak (fraction 1, Fig. 2). On this medium apoE is recovered in fractions 2 and 3, almost all apoC-I and considerable apoC-III-1 and apoC-III-2 in fraction 4, and all of the apoC-II and some apoC-III in fraction 5 (Fig. 2). This system is particularly useful for isolating apoC-II in preparation for ion-exchange chromatography, but the resolution of apoE is suboptimal.

ApoVLDL can additionally be resolved in a variety of chromatographic media eluted with guanidine-HCl or urea-containing buffers. The dissociating properties of guanidine are superior to those of urea, but guanidine is considerably more expensive. The C-apoproteins have apparent molecular weights in urea solutions of about 25,000 (10). In guanidine HCl they behave as proteins of 6-10,000 MW. The chromatographic profile (Fig. 3) is similar, but not identical, to that obtained with SDS. The void volume peak contains apoB and some intermediate-molecular weight proteins (fraction 1, Fig. 3), even when media such as Bio-Gel A 5M or Sepharose 6B (13) are employed. ApoE and other large proteins elute over a broad area (fractions 2 and 3, Fig. 3) and apoproteins C-II and C-III in

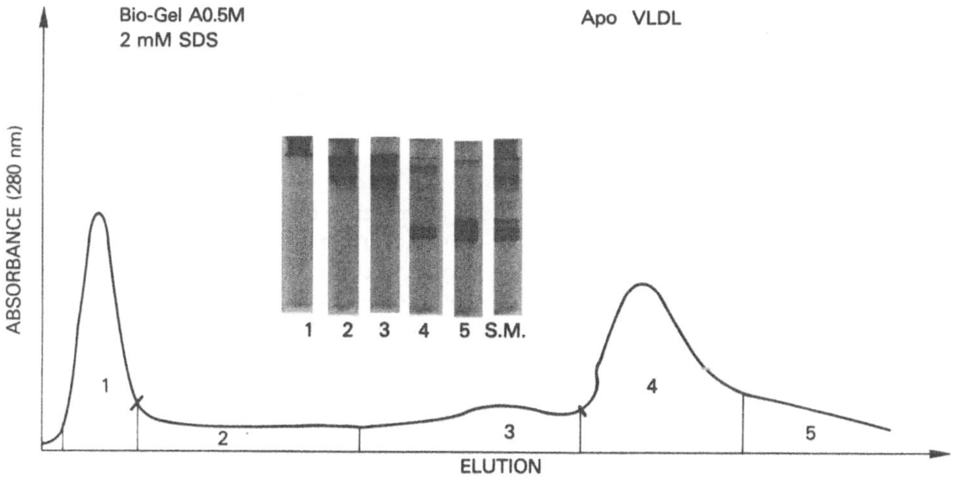

Fig. 2. Chromatography of apoVLDL on a 2.5 x 150 cm column of Bio-Gel A 0.5M. Conditions of polyacrylamide gel electrophoresis were as indicated in the legend to Fig. 1.

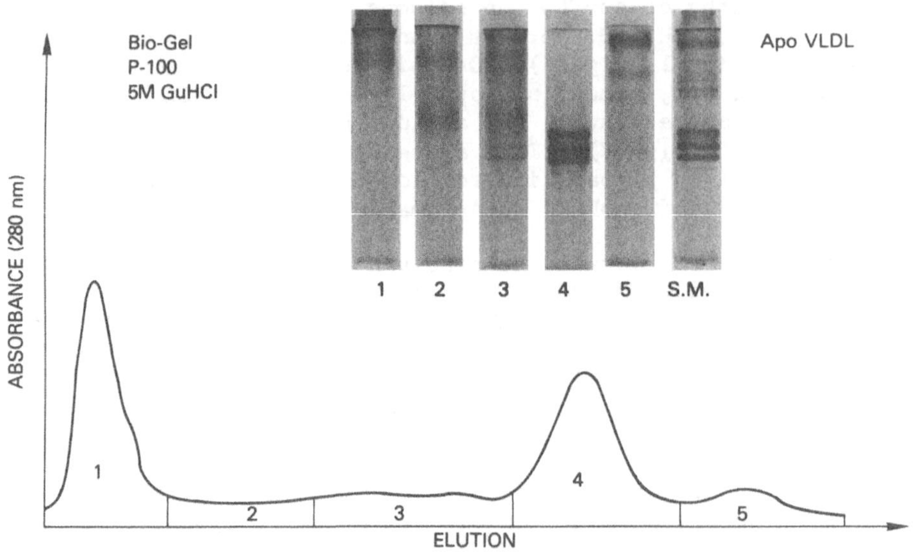

Fig. 3. ApoVLDL fractionated on a 2.5 x 140 cm column of Bio-Gel P-100 eluted with 0.1M NH_4HCO_3, 5M guanidine-HCl.

the second major peak (fraction 4, Fig. 3). ApoC-I elutes in a fairly well-defined peak (fraction 5, Fig. 3) after the other C-apoproteins. The late elution of apoC-I is attributable to the dissociating power of guanidine and the low molecular weight of this apoprotein (∿ 7,000).

Gel chromatography, therefore, permits rough quantitation of the B and C apoproteins in VLDL. The accuracy is perhaps better when SDS buffers are used in apoB estimations and urea or guanidine-HCl for the C-apoproteins. The C-apoproteins can be further fractionated by ion-exchange chromatography on DEAE-cellulose. DEAE-Sephadex does not generate peaks as sharp or as reproducible as the cellulose-based exchange medium. Shore and Shore have applied whole apoVLDL to DEAE-cellulose and resolved six to fifteen fractions (14, 15). We have rarely been able to obtain homogeneous fractions, however, unless preliminary gel chromatography is used to separate the C-apoproteins from the rest of the apoVLDL. The proteins of intermediate molecular weight (fractions 2 and 3, Figs. 1-3) elute from DEAE over a wide range of buffer conductivities and contaminate several of the C-apoprotein fractions. DEAE chromatography without urea does not permit good resolution of the C-II and C-III apoproteins (11). A gradient of Tris-HCl is better than one of NaCl. While good separation of the C-apoproteins is possible (10), recoveries from DEAE are usually not greater than 75 or 80%. It is not known whether the losses are selective and DEAE chromatography has, therefore, limited use in apoprotein quantitation. Apoproteins B and E are recovered in very low yields from DEAE and ion-exchange chromatography is not suited to quantitation or purification of these VLDL apoproteins.

ApoHDL. Gel chromatography on columns eluted with 6-8M urea have long been used to obtain estimates of the major protein components of apoHDL (7). Chromatography in acetic acid has yielded comparable results (8). The major HDL apoproteins, apoA-I and apoA-II, elute in distinct but incompletely separated peaks (Fig. 4). The void volume fraction (1, Fig. 4) is invariably present when acetic acid is used for chromatography and is rarely seen when the protein is solubilized and eluted with guanidine-containing buffers (Fig. 5). This peak consists mostly of irreversibly aggregated protein (both apoA-I and apoA-II). The first major peak (fraction 2, Fig. 4; and fraction 1, Fig. 5) contains the A-I apoprotein that

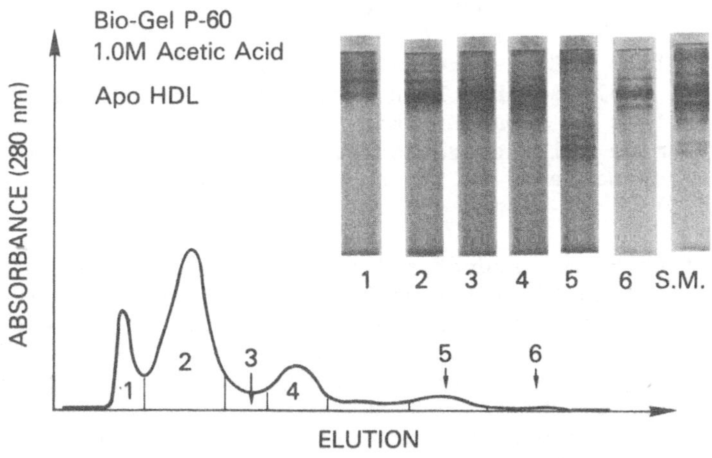

Fig. 4. Chromatography of apoHDL on a 2.5 x 140 cm column of Bio-Gel P-60 eluted with 1.0M acetic acid.

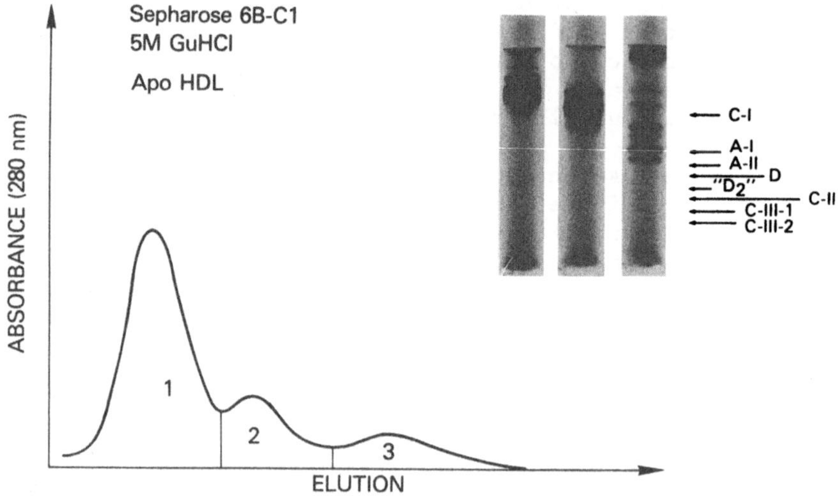

Fig. 5. Fractionation of apoHDL on a 2.5 x 140 cm column of Sepharose 6B-Cl eluted with 5M guanidine HCl.

usually is contaminated with apoA-II and occasionally with albumin and quantitatively minor HDL apoproteins. The valley between fractions 2 and 4 (Fig. 4) contains apoD (16, 17) and the second major peak (fraction 4, Fig. 4; and fraction 2, Fig. 5), the A-II apoprotein. The latter contains no tryptophan and consequently is not well-represented by the absorbance at 280 nm. The C-apoproteins elute next (fraction 5, Fig. 4; and fraction 3, Fig. 5). The C-apoprotein peak from HDL also contains the apoprotein recently described by Lim et al. (18) that has been tentatively designated "D-2". This latter apoprotein is apparent when most apoHDL samples are analyzed by polyacrylamide gel electrophoresis, but there is considerable variation among patients in the quantity present in apoHDL (cf., Figs. 4 and 5).

Several problems encountered in gel chromatography of apoHDL render it less than ideal for apoprotein quantitation. (1) Significant amounts of protein aggregate as a result of delipidation or incomplete dissociation during resolubilization. Protein aggregates elute in a void volume fraction that may account for variable (2-10%) proportions of the total recovered from chromatography. (2) In most, if not all, chromatographic systems, the A-I and A-II apoproteins are not completely separated. Moreover, both of the peaks comprising these apoproteins typically contain apoD. (3) ApoA-I often is detectable in all column fractions, particularly if buffer systems without guanidine-HCl are used. This phenomenon is particularly troublesome if acetic acid is employed. A distinct peak containing apoA-I elutes even near the salt volume (fraction 6, Fig. 4) where concentrated acetic acid used to solubilize the apoHDL emerges from the column. The size of this peak increases with succeeding column runs, suggesting resolubilization of precipitated apoA-I by other HDL apoproteins and the concentrated acetic acid.

ApoHDL can also be fractionated by DEAE chromatography. Recoveries, however, are less than quantitative, the C-apoproteins are not eluted in predictable positions, the A-I apoprotein is recovered in two or more peaks (19) over a broad range of conductivities (probably reflecting several states of aggregation), and the purity of fractions is variable.

Quantitation by Polyacrylamide Gel Electrophoresis

Estimates of the apolipoprotein content of isolated lipoprotein fractions can be obtained by polyacrylamide gel electrophoresis. The values obtained in most circumstances are relative, not absolute. The procedure can be employed to monitor changes in apoprotein content with dietary or drug perturbations or it can be employed in semi-quantitative comparisons of apoprotein patterns among different subjects.

The tetramethylurea (TMU) extraction procedure of Kane (6) has been used by several laboratories in combination with photodensitometry to analyze the protein content of apoVLDL and apoHDL. The great advantage of the procedure resides in its application to small quantities of lipoprotein. TMU appears to dissociate the lipoprotein lipid and selectively solubilizes all proteins except apoB (16). It has been used for both estimations of relative proportions of apolipoproteins (20) and in efforts to absolutely quantify the apoprotein constituents of VLDL (21). The procedure has been described in considerable detail (6, 21) and will not be recounted here. The use of such assays for apoprotein quantitation demands extensive validation and rigid control of experimental conditions. Apoproteins in pure form and in naturally occurring mixtures must be shown to produce a linear response in densitometric area. The limitations of sensitivity must be defined and the identity (and homogeneity) of stained bands confirmed by amino acid analysis and immunochemical techniques. Comparison of mobilities of known apoproteins with those in complex mixtures is not sufficient to establish identity. When TMU is used for lipoprotein delipidation, the influence of temperature on the recovery of soluble apoproteins must be considered (21). The volume of the stacking gels should be at least twice that of the largest sample to insure that the proteins are well-stacked before they enter the separating gel. ApoC-I does not stack well in most alkaline gel systems because of its high isoelectric point, and quantitation of this apoprotein is probably best performed in acid systems. Gel concentrations and buffer pH must be standardized because of their obvious influence on R_f, zone width and densitometric area. The disproportionate quantities of apoproteins in mixtures will often require electrophoresis at more than one load to obtain reproducible quantitation. The absolute chromogenicities of different apoproteins are not the same, and must be determined for the particular dye and staining conditions employed.

The influence of sample aging on the polyacrylamide gel electrophoresis pattern is one of the most difficult factors to control. An extreme example of alteration of apoVLDL patterns is displayed in Fig. 6. Comparison of the apoVLDL samples before fractionation (S.M.) demonstrates that the intensity of the C-II band was greatly diminished after storage of the plasma for one year at 4°C. The

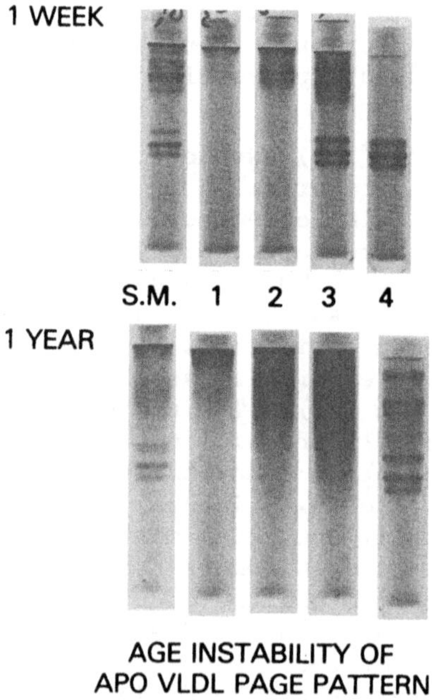

Fig. 6. Alteration of the polyacrylamide gel electrophoresis patterns of apoVLDL and its gel chromatographic fractions after storage of whole plasma for one year at 4°C. The apoVLDL shown in the upper panel was fractionated on a column of Bio-Gel A 0.5M (Fig. 2), that in the lower panel on Sepharose 6B-Cl eluted with 5M guanidine-HCl.

intermediate molecular weight apoproteins (fraction 2, Fig. 6) present distinct bands after one week of storage but a diffuse haze after one year. This appearance is typical of denatured and aggregated proteins. Analysis of residual phospholipid in fraction 2 (Fig. 6) after one year shows quantities as high as 10-15% of the protein weight after organic solvent extraction. It is possible that peroxidation of unsaturated fatty acids and subsequent formation of covalent bands with the apoprotein accounts for this phenomenon. Inspection of fraction 4 (Fig. 6) shows that only a trace of apoC-II band remains at one year, and there has appeared an intense band, absent initially, with mobility distinctly less than apoC-II. Kane and coworkers (21), moreover, have noted that the form of apoC-III of greatest mobility is diminished with storage, and we have noted the appearance of several new, uncharacterized bands.

The capriciousness of column chromatography and polyacrylamide gel electrophoresis in the quantitation of apoproteins in isolated lipoproteins has been emphasized not to denigrate their usefulness, but to underscore the limitations inherent in their use. They have been widely employed in many laboratories including that of the authors, but they will probably be displaced in time by immunochemical methods that are more readily standardized and which require less sample manipulation.

TECHNIQUES APPLICABLE TO UNFRACTIONATED PLASMA

The specificity and sensitivity of several immunochemical methods render them ideal for apolipoprotein quantitation in unfractionated plasma or serum in addition to isolated lipoprotein fractions. All are based on the principle of titration - the reactivities of test samples are matched with those of known standards. The choice of techniques depends on several factors, including the concentration of the apoprotein in plasma (Table 1), the number of samples to be processed, the degree of precision and accuracy required, and the availability or expense of necessary equipment and reagents. Pure antigens are mandatory for most assay systems with the exception of two-dimensional immunoelectrophoresis which will not be discussed. Monospecific antisera are necessary for some, but not all, methods.

Purification of Antigens. The methods of apoprotein fractionation by gel and ion-exchange chromatography discussed above provide general outlines of the most widely used techniques of apoprotein purification. It is a common experience, however, that two chromatography steps are frequently inadequate to produce antigens of sufficient homogeneity that truly monospecific antisera can be produced. We will briefly sketch here our own approaches for the production of good antigens, recognizing that methods used by others may be comparable or superior.

Table 1. Approximate* Quantities of Apolipoproteins in Normal Plasma (mg/dl)

A-I	80-120
A-II	30-50
B	70-90
C-I	3-7
C-II	3-5
C-III	8-12
D	2-4
E	3-6
"D-2"	1-2

*Reliable estimates for several of these proteins are not available. The values presented for these are only educated guesses.

HDL to be used for apoA-I isolation is prepared by adjusting fresh plasma to 1.090 g/ml with KBr and separating the VLDL, LDL, and Lp(a) lipoproteins by ultracentrifugation. The infranatant fraction is adjusted to 1.21 g/ml and the HDL isolated by a second ultracentrifugation step. The density of the HDL is raised to 1.23 g/ml with solid KBr, an aqueous solution of KBr at 1.21 g/ml is layered over it, and ultracentrifugation repeated to remove most of the serum albumin. The HDL are dialyzed and totally delipidated with the methanol-diethylether procedure described above. ApoHDL is resolubilized in 5M guanidine HCl and fractionated on columns of Bio-Gel A 0.5M, P-100 or Sephacryl. The latter may provide optimal results. The several tubes around the peak of the apoA-I fraction are pooled, dialyzed against 5mM NH_4HCO_3 and lyophilized. This crude apoA-I is further purified by DEAE-cellulose chromatography in 6M urea using a linear gradient from 0.05M to 0.15M Tris-HCl. The apoA-I elutes in two or three poorly separated peaks, each of which is individually pooled and analyzed for homogeneity by alkaline and SDS polyacrylamide gel electrophoresis and amino acid analysis. Only apoA-I that is pure by all criteria is used to raise antisera.

HDL for apoA-II isolation is prepared and delipidated as described for apoA-I. However, whole apoHDL is first applied to DEAE-cellulose and the crude apoA-II fraction pooled, dialyzed and lyophilized. Final purification is achieved by chromatography in guanidine HCl on columns of Bio-Gel P-100 or Sephacryl. If polyacrylamide gel electrophoresis or amino acid analysis suggest contamination, the apoA-II should be reduced, carboxymethylated and subjected to a second gel chromatography step before use as antigen.

Antisera reactive with lipid-free B protein are not easily produced by immunization of animals with whole LDL. Therefore, apoB is first produced in homogeneous form. LDL is isolated in a narrow density range, 1.035 < d < 1.050 g/ml, from fresh plasma or sera. The LDL is recentrifuged once at 1.050 g/ml, dialyzed, and delipidated with the methanol-diethylether solvents. Before drying under nitrogen, solid SDS sufficient for a concentration of 0.2M in a protein solution of 10 mg/ml is added. The apoB is fractionated on a column of Bio-Gel A 0.5M or A 1.5M eluted with 0.05M NH_4HCO_3, 2 mM SDS. The tubes on either side of the void volume peak are pooled and lyophilized, those on the ascending and descending limbs are discarded. The SDS in the column effluent permits ready solubilization of the lyophilized apoB in 0.1M NH_4OH, and the resolubilized apoB is subjected to a second purification step on the same column before use as antigen.

ApoC-I for immunization is obtained in the break-through volume of the DEAE-cellulose columns used to fractionate the C-apoproteins of apoVLDL (10). Before use as antigen, the apoC-I is purified by gel chromatography in guanidine HCl on columns of Sephadex G-75, Bio-Gel P-60 or Bio-Gel P-100. In spite of its low molecular weight, apoC-I is very immunogenic.

ApoC-II is the most difficult of the C-apoproteins to obtain in a high degree of purity. Chromatography of apoVLDL on Bio-Gel A 0.5M (Fig. 2) is most useful for this purpose because apoC-II is recovered near the end of the chromatogram free of contamination with intermediate-molecular weight proteins. When apoC-II is to be used for immunization, the appropriate peak from DEAE-cellulose (10) should be cut very narrowly to minimize the possibility of contamination with apoC-III. The latter cannot be separated from apoC-II by gel chromatography since their molecular weights are almost identical. Antisera raised to apoC-II that react with apoC-III can be easily adsorbed, since apoC-III is easy to prepare in homogeneous form.

ApoC-III for immunization is obtained as described (10). ApoC-III-1 is best for this purpose, and only peak tubes from DEAE-cellulose (10) should be used as antigen. The apoC-III-2 peak is frequently contaminated with proteins of similar electrophoretic mobility and size. Antisera raised to apoC-III-2 are often oligospecific.

The authors have not succeeded in purifying apoD. Published methods for apoD isolation (16, 17) detail the difficulties usually encountered. Fortunately, many antisera raised to crude apoA-II react with apoD and specific anti-apoD can be obtained by adsorption.

ApoE, like apoD, is purified with considerable difficulty. A crude preparation is obtained by chromatography of apoVLDL (ideally from a patient with type 3 hyperlipoproteinemia) on Bio-Gel A 0.5M (SDS buffer system) and rechromatography on Sephacryl or Bio-Gel A

0.5M or 1.5M eluted with 5M guanidine HCl. Preparative SDS polyacrylamide gel electrophoresis is the final purification step. This can be accomplished using a commercial preparative gel apparatus (22). Alternatively, the apoprotein E zone (identified by use of a fixed and stained marker gel) can be sliced from analytical calibre gels, the gels homogenized in 0.1M NH_4OH, and the gel slurry separated by low speed centrifugation. If the slurry is extracted a second time with 0.1M NH_4HCO_3, about 60% of the apoE can be recovered. After dialysis and lyophilization, a preparation suitable for immunization is obtained. Alternatively, the minced gel slices can be used directly for immunization.

Preparation of Antisera. One to three mg of pure apoprotein is usually sufficient to raise moderate titer antisera in rabbits or other small rodents. Useful antisera have been obtained, however, with a total of 0.5 mg protein. Larger quantities, but not more than 6 or 8 mg, are sometimes needed for sheep or goats. Rabbits can be injected in foot pads or (more humanely) intradermally on the back. The apoprotein solution is thoroughly emulsified in complete Freund's adjuvant. One-fourth of the total available dose is administered each injection. A minimum of three injections at biweekly intervals is necessary and higher-titer antisera can be produced if sufficient material is available for 4-6 injections. The use of larger amounts of protein at less frequent intervals usually induces lower titer antisera. If the immunochemical assay is to be used for a considerable period of time, it is useful to pool several bleeds from rodents or at least two from sheep or goats to avoid repeated tedious characterization and standardization of antisera.

It is often desirable to optically clarify antisera or to remove plasma proteins that interfere with immunochemical assays. Antisera produced in rabbits may be grossly lipemic, a reflection of even mild infection at injection sites. Lipoproteins can be separated from the gammaglobulins and other serum proteins by a single ultracentrifugation at 1.21 g/ml. The infranatant fraction is dialyzed against 0.2M NaCl, 0.05M NH_4HCO_3, filtered through glass wool, bacteriostatic agents are added, and the antisera frozen until use. Hemolyzed antisera raised in sheep can be clarified and the gammaglobulin fraction isolated by DEAE-cellulose chromatography. The column is equilibrated with 0.03M Tris-HCl, pH 8.0, and the antisera dialyzed against the same buffer. Hemoglobin and most of the plasma proteins are bound to the DEAE and the gammaglobulins can be eluted with the equilibration buffer and lyophilized. Alternatively, the gammaglobulins in crude form can be precipitated from the antisera by mixing with an equal volume of 28% (w/v) Na_2SO_4, and washing the precipitate with 14% Na_2SO_4. Lipoproteins are precipitated under these conditions, producing turbidity in the resolubilized gammaglobulin solution. A simple method avoiding precipitation of lipoproteins is that employing caprylic acid for isolation of mammalian IgG (24).

Antisera or purified gammaglobulins are stored frozen in small aliquots to avoid repeated freezing and thawing. A satisfactory bacteriostatic agent is a mixture of $3.45 \times 10^{-3}M$ NaN_3 and $1.5 \times 10^{-4}M$ thimerosal (1:1,000, v:v). Antisera stored at 4°C for prolonged periods usually lose activity.

Gel Immunodiffusion

Quantitation of plasma proteins by a variety of immunodiffusion techniques has been used for 20 years. Agar and agarose have widely been chosen as the stabilizing media, although cellulose acetate polyacrylamide, filter paper, and other media have found occasional use. The liquid phase usually contains 0.15 - 0.20M NaCl and an appropriate buffer to maintain pH at 7.0-8.5. At pH > 9.0, antigen-antibody complexes may be dissolved and many antigens are insoluble at more acid pH. When chicken antibodies are used, a gel containing 10% NaCl has been recommended. We routinely use a 0.05M barbital (Veronal) buffer of pH 8.2 and an agar or agarose final concentration of 1%. In preparation for quantitative immunodiffusion, it is convenient to prepare a 2% gel solution by boiling the agar or agarose in the liquid phase containing 1:10,000 thimerosal (w:v), filtering the solution through a hot glass funnel with cheese cloth or glass wool, and storing the gel at 4°C in 25 ml or 50 ml Ehrlenmeyer flasks covered with parafilm. Plates for single immunodiffusion are prepared by melting the gel in a boiling water bath. The gel is cooled to 56°C (before mixing with an equal volume of antiserum), diluted in the liquid phase, and heated to 56°C. A 1-2 mm layer is poured on a glass plate or petri dish (23). The dilution of antiserum that is appropriate must be determined empirically and is highly contingent on the antibody titer. Dilutions in a range of 1:10 to 1:1,000 may be appropriate. When only small dilutions are possible, the procedure may prove too expensive in terms of precious antisera. Techniques are available, however, for increasing the sensitivity (24). The titers of antisera raised to many of the apolipoproteins are often much lower than those of commercial antisera raised to abundant proteins. This is usually secondary to the limited quantities of homogeneous antigens available for injection and the impossibility of immunizing several animals simultaneously.

Specific antisera are generally required to use the single radial immunodiffusion (RID) method for protein concentration. Polyspecific antisera will generate multiple precipitin rings, and the correlation of individual precipitin rings with specific antigens can be a cumbersome undertaking (25). Dilutions of antigen of known concentration are used to construct the standard curve, and each point is determined in duplicate on different areas of the plate. In addition, a standard preparation of lipoprotein solution, serum or plasma should be used with every assay to control for interassay variation.

The preparation of test samples for analysis requires special attention, particularly when purified apoproteins serve as standards and lipoproteins or serum are to be assayed. The reactivity of antigens that are complexed in one or more lipoproteins should be considered unknown, and a variety of pretreatment procedures evaluated. Total delipidation of lipoproteins or plasma with organic solvents is often considered optimal to insure maximum reactivity. Treatment with organic solvents, however, is time-consuming and complicated by uncertainty concerning selective or non-specific losses, and alteration of physical state (e.g., aggregation) and thereby immunoreactivity of the apoproteins. The use of ionic or non-ionic detergents, or dissociating agents such as guanidine and urea, to quantitatively unmask antigenic sites has been disappointing. Albers and coworkers (26) have treated plasma and HDL samples with TMU and a Tris buffer containing 8M urea, and have apparently achieved adequate exposure of the antigenic sites of apoA-I. TMU treatment, therefore, appears promising as an agent to maximize immunoreactivity of apolipoproteins and to minimize the handling of samples before actual assay.

The theoretical basis of quantitation in RID is still not well-defined. In general, the diameter of the precipitin ring is directly related to the quantity of antigen in the well and inversely related to the concentration of antibody in the gel. The reaction cannot be viewed as proceeding at equivalence, however, a fact visually confirmed by the non-uniformity of precipitate within the precipitin area (28). The usual response variables for RID have been the area, diameter, or square of diameter of the precipitin ring. A linear relationship of dose to squared diameter (29) or area (30) is expected. The appropriate duration of incubation before measurement of the precipitin ring is determined empirically. The choice is dictated by considerations of convenience as well as the characteristics of the systems. Readings may be taken as early as 24 hours in some cases (29), although the reaction may proceed for 14 days or more (27). Impedence to diffusion of large macromolecules is known to artifactually reduce their apparent concentrations (28), a phenomenon that may limit the potential usefulness of RID for apoB quantitation. Similarly, apoproteins bound to lipoproteins cannot be expected to behave identically to their lipid-free counterparts.

The limit of sensitivity of most RID systems is on the order of 1 mg/dl. RID is theoretically well-suited for the quantitation of apolipoproteins A-I, A-II and B. ApoD, apoE and the C-apoproteins are present in undiluted plasma at concentrations near the practical lower limit of RID analyses. Minor modifications of the classical method (24), however, may even permit their quantitation.

RID, like all immunoassays, requires very careful attention to technical details and quality control. Errors can arise and be compounded by inaccurate dilution or application of test samples, variations of gel thickness within a plate, incomplete mixing of antiserum and gel, inadequate spacing of test wells, dehydration of gel, etc. Nevertheless, the method requires a minimum amount of specialized equipment and technical expertise, reagents are inexpensive, and a large number of samples can be economically processed.

Electroimmunoassay

Two quantitative immunoelectrophoretic methods have been applied to the study of apolipoproteins. Bidimensional crossed immunoelectrophoresis potentially can be used with polyspecific antisera and may permit simultaneous quantitation of more than one antigen in a complex mixture. Its use is being evaluated in several laboratories, but as yet there are no published examples of its application to apolipoprotein quantitation. The technique is technically demanding and not readily applied to large numbers of samples. A useful literature review of this and other immunoelectrophoretic methods has been published by Verbruggen (31) and this should be consulted for a general summary of theoretical and practical considerations.

Electroimmunoassay (EIA) has been used to quantitate the proteins of both alpha- (32-35) and betalipoproteins (36-38). A monospecific antisera is incorporated in an agarose gel as described above for RID. Antigen is added to wells punched in the gel and electrophoresis results in the production of rocket or spiked shaped precipitins, the height or area of which is proportional to antigen concentration (39, 40). The conditions of electrophoresis are chosen so that antibody is presumed to be stationary (31). The anodally migrating antigens precipitate with antibodies but unbound antigen within the precipitin cones migrates into and continues to redissolve the precipitate, displacing the precipitin zones toward the anode, until free antigen is finally exhausted.

In EIA, as in RID, the assay conditions must be determined empirically. Conditions are sought wherein the smallest amount of antigen in the lowest possible antiserum concentration gives a well-defined precipitin peak of 2-5 cm in height. The useful antiserum concentration may be as low as 0.5% or as high as 20%. The minimum amount of antigen that can be confidently quantitated may range from 3-100 ng. The incorporation in the gel of so-called "enhancers" of immune precipitation, such as dextrans or polyethylene glycol, can augment the sensitivity by a factor of 2-5 (41, 42). Using Dextran T10 (Pharmacia Fine Chemicals), Curry _et al_. were able to measure 20 ng of apoA-I or apoA-II in EIA.

The factors which can affect quantitation with EIA and their evaluation are discussed elsewhere (31) and will not be elaborated here. The results to date with apolipoproteins have been puzzling and at variance with much other published work. Durrington and co-workers (38) discovered that response curves of LDL and serum were not parallel in EIA and a secondary standardization was necessary to obtain quantitation. Farish et al. (34) obtained values for "apolipoprotein A" that are higher than those predicted from ultra-centrifugation (1) or radioimmunoassay (vide infra). Still higher "apoA" levels were reported by Curry et al. (35), who, moreover, found apoA-I/apoA-II weight ratios considerably different from those estimated by other techniques. In neither of these studies was it deemed necessary to delipidate the plasma or lipoprotein solutions to unmask HDL antigenic sites, whereas such treatment has generally been required in other immunoassays. In addition, the possibility that isolated apolipoprotein "standards" were less reactive than the naturally occurring species in intact lipoproteins was not excluded. Both studies suggested that "apoA" could be quantitated directly by EIA using a polyspecific antiserum to HDL. This impression is conceptually misleading. The antisera employed undoubtedly contained antibodies recognizing both the A-I and A-II apoproteins, but the EIA theory would predict that only one group of antibodies (probably that present in highest titer) determined the EIA result. The "apo-A" quantitation obtained, therefore, would be valid only if the test sample had the same (A-I/A-II) and (A-I+A-II/total HDL apoprotein) ratios as the "standard" apoHDL. The results presented in one paper (35) suggest that such an assumption may not be justified.

EIA has been recommended for protein quantitation because of its high sensitivity, conservation of antisera, and rapidity on comparison with diffusion techniques. However, it is not a simple method to execute, standardize, or quality control, and it is not applicable to large scale studies. Extensive validation is mandatory before quantitation with this method can be accepted without qualification.

Radioimmunoassay

The extreme sensitivity of radioimmunoassay (RIA) is not required for quantitation of most of the known plasma apolipoproteins (see Table), but it is ideal for studies of apoprotein metabolism in cell suspensions, tissue culture and isolated perfused organs. The high degree of automation possible with this technique, moreover, renders it particularly suitable for the analysis of large numbers of samples. Of the techniques discussed above, only radial immunodiffusion can be comparably adapted to mass screening efforts.

The basic elements of RIA include: (1) antibody; (2) labeled and unlabeled antigen; (3) test sample; (4) incubation; (5) separation of antibody-bound and free antigen; (6) radioactivity measurement; and (7) data reduction. Items four through seven present no problems peculiar to apolipoproteins and are treated in detail in most monographs and reviews of RIA. The first three elements warrant consideration here for reasons apparent in the discussion of other immunoassays.

Antisera used in RIA generally need not be specific for a single antigen except in the unusual circumstance where the antigen to be labeled is not pure. Grossly impure antigens are of little value in any event because of inevitable difficulties in establishing the immunoreactivity of the antigen after radiolabeling. It is desirable that the antibodies be directed against regions of the apolipoprotein molecule that are fully exposed in native lipoproteins. This circumstance would minimize or obviate the delipidation procedures necessary before actual assay. Theoretically, such antisera might be produced if native lipoproteins are employed for immunization. Most of the RIAs that have been developed for apoB have employed antisera raised to LDL (43-46) and there has been no need to treat test samples with detergents or organic solvents. However, apoB may represent a special case. An RIA based on an antiserum to lipid-free and chromatographically purified apoB has been reported (47). Pretreatment of test samples was unnecessary, but apoB in delipidated lipoproteins was only about 15% as reactive as native LDL (47). The standard antigen in all these assays was LDL, not apoB, and the poor reactivity of lipid-free B protein was ascribed without proof to aggregation (47). There exists additional uncertainty concerning the potential of RIA with LDL as standard to quantitate the apoB content of VLDL (44, 45, 47).

The possibility of using antisera to native lipoproteins to quantitate other apolipoproteins has not been fully evaluated. The antisera employed for RIA of apoA-I (48-51) and apoA-II (51, 52) have been raised to purified apoproteins with the exception of that reported by Fainaru _et al_. (49) in which antisera to both HDL and apoA-I were used. In all assays except the latter, pretreatment of test samples was required. Schonfeld and Pfleger (48) found that less than 30% of plasma apoA-I was recognized in RIA without organic solvent delipidation. Treatment with urea and guanidine, or freeze-thawing produced suboptimal quantitation. Karlin and coworkers (50) were able to expose the apoA-I antigenic sites in plasma by the simple maneuver of diluting and heating. Only Fainaru _et al_. (49) have claimed that plasma apoA-I could be quantitated without pretreatment of test samples, and this was possible in their assay system using antibodies raised to HDL, apoHDL, and apoA-I. It is unlikely, therefore, that the specificities of the antisera account for the differences between the assay of Fainaru (49) and those of Schonfeld and Karlin. Differences in antibody specificity, however, may account for the relative ease with which Karlin _et al_. exposed the immunoreactive sites of apoA-I in their assay (50).

Schonfeld *et al.* (53) have demonstrated that an anti-apoA-I serum in their laboratory contains at least two subpopulations of antibodies specific for the carboxyl and amino-terminal of apoA-I and without cross-reactivity. One of these antibody subpopulations was more reactive with native HDL (53).

If VLDL or HDL are used to produce antisera to their apolipoproteins, it will probably be necessary to purify the antisera by immunoadsorption before use in RIA. This is because of the apolipoprotein heterogeneity of these classes of lipoproteins and the rapidity with which these lipoproteins exchange. Thus, if an antiserum used in an RIA for apoC-III also precipitated apoB containing lipoproteins, then the presence of C-III and B on the same lipoprotein could lead to "nonspecific" precipitation of C-III by anti-apoB.

Radiolabeling of apoproteins has generally been performed chemically with chloramine-T (54) or iodine monochloride (55), or enzymatically with lactoperoxidase (56). The chloramine-T procedure has been satisfactory for isolated apoproteins, but one of the other procedures is preferable when labeled lipoproteins are to be used as standards. Lipoproteins from animals particularly may contain an abundance of polyunsaturated fatty acids and a large amount of label may be incorporated in the lipid when chloramine-T is used (57).

The high degree of sample dilution necessary to conserve antisera in the RIA of apolipoproteins is one of the major drawbacks of this method. Automation of the procedure is also relatively expensive and the work of pretreating samples is tedious and time-consuming. (Attempts to use TMU for this purpose have not been published). The sensitivity and accuracy of the technique, however, hold great appeal in innumerable research situations and increasing use of RIA and other immunoassays can be anticipated in studies of lipoprotein physiology.

REFERENCES

1. Havel, R.J., Eder, H.A., and Bragdon, S.H. J. Clin. Invest. 34, 1345-1353, 1955
2. Herbert, P.N., Forte, T.M., Shulman, R.S., LaPiana, M.J., Gong, E.L., Levy, R.I., Fredrickson, D.S., and Nichols, A.V. Prep. Biochem. 5, 93-129, 1975
3. Scanu, A.M., and Edelstein,C. Anal. Biochem. 44, 576-588, 1971.
4. Folch, J., Lees, M., and Sloane-Stanley, G.H. J. Biol. Chem. 226, 497-509, 1957
5. Helenius, A., and Simons, K. Biochem. 10, 2542-2547, 1971.
6. Kane, J.P. Anal. Biochem. 53, 350-364, 1973
7. Scanu, A., Toth, J., Edelstein, C., Koga, S., and Stiller, E.

Biochem. 8, 3309-3316, 1969
8. Rudman, D., Garcia, L.A., and Howard, C.H. J. Clin. Invest. 49, 365-372, 1970
9. Scanu, A.M., Lim, C.T., and Edelstein, C. J. Biol. Chem. 247, 5850-5855, 1972
10. Herbert, P.N., Shulman, R.S., Levy, R.I., and Fredrickson, D.S. J. Biol. Chem. 248, 4941-4946, 1973
11. Brown, W.V., Levy, R.I., and Fredrickson, D.S. J. Biol. Chem. 244, 5687-5694, 1969
12. Reisfeld, R.A., and Small, P.A., Jr. Science 152, 1253-1255, 1966
13. Shelburne, F.A., and Quarfordt, S.H. J. Biol. Chem. 249, 1428-1433, 1974
14. Shore, B., and Shore, V. Biochem. 8, 4510-4516, 1969
15. Shore, V.G., and Shore, B. Biochem. 12, 502-507, 1973
16. McConathy, W.J., and Alaupovic, P. FEBS Letts. 37, 178-182, 1973
17. Kostner, G.M. Biochim. Biophys. Acta 336, 383-395, 1974
18. Lim, C.T., Chung, J., Kayden, H.J., and Scanu, A.M. Biochim. Biophys. Acta 420, 332-341, 1976
19. Lux, S. E., and John, K. M. Biochim. Biophys. Acta 278, 266-270, 1972
20. Carlson, L.A., and Ballantyne, D. Atherosclerosis 23, 563-568, 1976
21. Kane, J.P., Sata, T., Hamilton, R.L., and Havel, R.J. J. Clin. Invest. 56, 1622-1634, 1975
22. Utermann, G. Hoppe-Seyler's Z. Physiol. Chem. 356, 1113-1121, 1975
23. Steinbuch, M., and Andran, R. Arch. Biochem. Biophys. 134, 279-284, 1969
24. Albers, J.J., and Hazzard, W.R. Lipids 9, 15-26, 1974
25. Darcy, D.A. Clin. Chim. Acta 38, 329-337, 1972
26. Mancini, G., Nash, D.R., and Heremans, J.F. Immunochem. 7, 261-264, 1970
27. Albers, J.J., Wahl, P.W., Cabana, V.G., Hazzard, W.R., and Hoover, J.J. Metabol. 25, 633-644, 1976
28. Trautman, R., Cowan, K.M., and Wagner, G.G. Immunochem. 8, 901-916, 1971
29. Fahey, J.L., and McKelvey, E.M. J. Immunol. 94, 84-90, 1965
30. Mancini, G., Carbonara, A.O., and Heremans, J.F. Immunochem. 2, 235-254, 1965
31. Verbruggen, R. Clin. Chem. 21, 5-43, 1975
32. Johansson, B.G., and Laurall, C.-B. Scand. J. Clin. Lab. Invest. 23, 231-233, 1969
33. Johansson, B.G., and Medhus, A. Acta Med. Scand. 195, 273-277, 1974
34. Farish, E., Shepherd, J., Lawrie, T.D.V., and Morgan, H.G. Clin. Chim. Acta 62, 97-101, 1975
35. Curry, M.D., Alaupovic, P., and Suenram, C.A. Clin. Chem. 22, 315-322, 1976

36. Kahan, J., and Sundblad, L. Scand. J. Clin. Lab. Invest. 24, 61-68, 1969
37. Lasser, N. L., Roheim, P.S., Edelstein, D., and Eder, H.A. J. Lipid Res. 14, 1-8, 1973
38. Durrington, P.N., Whicher, J.T., Warren, C., Bolton, C.H., and Hartog, M. Clin. Chim. Acta 71, 95-108, 1976
39. Laurell, C.-B. Anal., Biochem. 15, 45-52, 1966
40. Laurell, C.-B. Scand. J. Clin. Lab. Invest. 29, Suppl. 124, 21-37, 1972
41. Harrington, J.C., Fenton, J.W., II, and Pert, J.H. Immunochem. 8, 413-421, 1971
42. Kostner, G., and Holasek, A. Anal. Biochem. 46, 680-683, 1972
43. Eaton, R.P., and Kipnis, D.M. J. Clin. Invest. 48, 1387-1396, 1969
44. Schonfeld, G., Lees, R.S., George, P.K., and Pfleger, B. J. Clin. Invest. 53, 1458-1467, 1974
45. Albers, J.J., Cabana, V.G., and Hazzard, W.R. Metab. 24, 1339-1351, 1975
46. Bautovich, G.J., Simons, L.A., Williams, P.F., and Turtle, J.R. Atherosclerosis 21, 217-234, 1975
47. Thompson, G.R., Birnbaumer, M.E., Levy, R.I., and Gotto, A.M., Jr. Atherosclerosis 24, 107-118, 1976
48. Schonfeld, G., and Pfleger, B. J. Clin. Invest. 54, 236-246, 1974
49. Fainaru, M., Glangeaud, M.C., and Eisenberg, S. Biochim. Biophys. Acta 386, 432-443, 1975
50. Karlin, J.B., Juhn, D.J., Starr, J.I., Scanu, A.M., and Rubenstein, A.H. J. Lipid Res. 17, 30-37, 1976
51. Assmann, G., Smootz, E., Adler, K., Capurso, A., and Oette, K. J. Clin. Invest. 59, 565-575, 1977
52. Mao, S.J.T., Gotto, A.M., Jr., and Jackson, R.L. Biochem. 14, 4127-4131, 1975
53. Schonfeld, G., Bradshaw, R.A., and Chen, J.-S. J. Biol. Chem. 251, 3921-3926, 1976
54. Roth, J. In O'Malley, B.W.O., and Hardman, J.G. (Eds.): Methods in Enzymology XXXVII. New York, Academic Press, 1975, pp. 223-233
55. McFarlane, A.S., Munro, H.N., and Allison, J.B. (Eds.: Metabolism of Plasma Proteins in Mammalian Protein Metabolism. Academic Press, New York, 1964, p. 331.
56. Marchalonis, J.J. Biochem. J. 113, 294-305, 1969
57. Fidge, N.H., and Poulis, P. Clin. Chim. Acta 52, 15-26, 1974

ISOLATION AND COMPOSITION OF HUMAN PLASMA APOLIPOPROTEINS

Henry J. Pownall
Baylor College of Medicine and the Methodist Hospital
Houston, Texas 77030, U.S.A.

I. INTRODUCTION

The biochemist's attempts to understand the structure and function of the human plasma lipoproteins have expanded greatly in recent years due to their possible role in the development of atherosclerosis. Five of the major apolipoproteins have been sequenced and our view of the structural organization of lipoproteins has seen considerable refinement. All of these achievements depend, in one way or another, upon the investigators skill in isolating both intact lipoproteins and apolipoproteins in a highly purified and homogeneous state. For this reason, we present in this chapter a description of some of the tried and proven methods for the purification of the lipoproteins and their component apolipoproteins and lipids. This presentation will not include the purification of abberant lipoproteins; the less frequently used or exotic techniques will be mentioned only to provide a reference for those who may be interested in the more specialized methods. Three reviews of lipoproteins provide ample source material on those areas of lipoprotein structure and function which are outside the scope of this volume (1-3).

II. COMPOSITION OF THE HUMAN PLASMA LIPOPROTEINS

A. Composition of Human Plasma High Density Lipoprotein (HDL)

The lipid-protein of HDL is depicted in Figure 1. The HDL (HDL_2 and HDL_3) are approximately 50% protein and 50% lipid. The protein composition is predominantly apoA-I (71%) and apoA-II (21%) with small amounts (8%) of apoC and apoD composing the remainder (4). The ratio of apoA-I to apoA-II was reported as 2:1 (5). The sequence of amino acids composing apoA-I and apoA-II are known (6,7) and are given in Figure 2 in terms of the one letter symbols for the amino acids (8). Phospholipids and cholesterol represent the major lipids of HDL; smaller amounts of cholesteryl esters and mono-, di-, and triglycerides account for the remainder. Phosphatidylcholine (75%) and sphingomyelin (13%) represent the dominant phospholipids which are accompanied by smaller quantities of other phospholipids and lysophosphatides (4).

The fatty acid composition of the individual lipids of the HDL from normal fasting subjects on an ad lib diet are shown in Table 1. Oleate, linoleate and palmitate represent the major fatty acids of the phospholipids and triglycerides whereas linoleate is the predominant fatty acid of the cholesteryl esters (9).

B. Composition of Human Plasma Low Density Lipoprotein (LDL)

The composition of plasma LDL from normal subjects is given in Figure 3. The particle contains 25% protein and 75% lipid. The major protein of LDL, apoB, represents its entire protein content. LDL contains 37% cholesteryl ester, 21% phospholipid, 9% cholesterol and 8% triglyceride. The major phospholipid classes are phosphatidylcholine (65%) and sphingomyelin (25%). The fatty acid composition of the individual lipid classes of LDL isolated from normal fasted subjects are given in Table 2; palmitate and oleate are the major fatty acids in the phospholipids of LDL (4). Cholesteryl esters of LDL are composed mainly of oleate and linoleate whereas the LDL triglycerides are composed largely of palmitate and oleate (4).

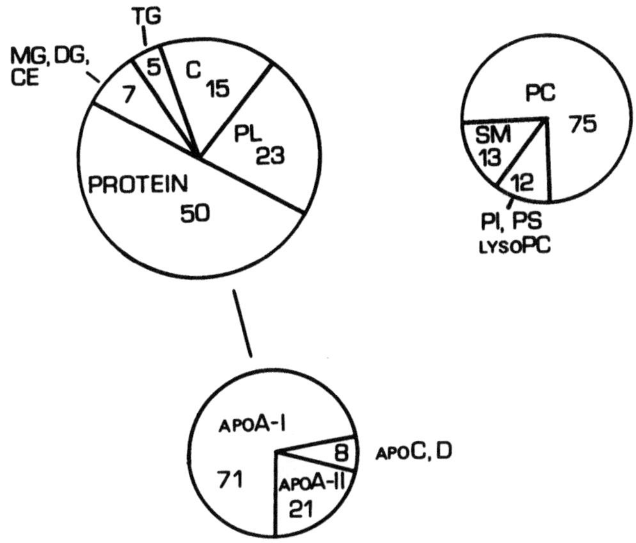

Figure 1: Lipid-protein composition of HDL.

apoA-I	DEPPQ	SPWDR	VKDLA	RVYVD	VLKDS	GRDYV	30
	SQFQG	SALGK	QLNLK	LLWDD	VTSTF	SKLRQ	60
	ELGPV	TEEWF	NDLQE	KLNLE	KETGE	LRQEM	90
	SKDLE	EVKAK	VQPYL	DDFQK	KWQEM	ELYRQ	120
	KEVPL	RAELQ	EGARQ	KLHEL	QEKLS	PLGEE	150
	MRDRA	RAHVD	ALRTH	LAPYS	DELRQ	RLAAR	180
	LEALK	ENGAG	RLAEY	HAKAT	EHLST	LSEAK	210
	KPALE	DLRQG	LLPVL	ESFKV	SFLSA	LEEYT	245
	KLNTQ						
apoA-II	ZAKEP	CVESL	VSQYF	QTVTD	YGKDL	MEKVK	30
	SPELQ	AQAKS	YFEKS	KEQLT	PLIKK	AGTEL	60
	VNFLS	YFVEL	GTQPA	TQ			77

Figure 2: Amino acid sequences of apoA-I and apoA-II monomer. The amino terminal residue of apoA-II is pyrrolidone carboxylic acid. ApoA-II occurs as a dimer of identical chains linked by a disulfide bridge at residue-6.

Table 1

Average Fatty Acid Composition of HDL_3 of Four Normal Subjects on an Ad Lib Diet

Fatty acid[a]	Phospholipids	Cholesteryl esters	Triglycerides
14:0	1.02	1.60	2.79
16:0	30.3	13.5	26.4
18:0	17.7	1.46	5.46
16:1	1.20	2.99	5.01
18:1	14.4	19.1	39.5
18:2	24.5	56.9	19.2
18:3,20:3	4.11	trace	trace
20:4	9.56	4.56	trace

[a] Adapted from Morrisett el al. (10). X:Y denotes X carbons in an acyl chain having Y double bonds.

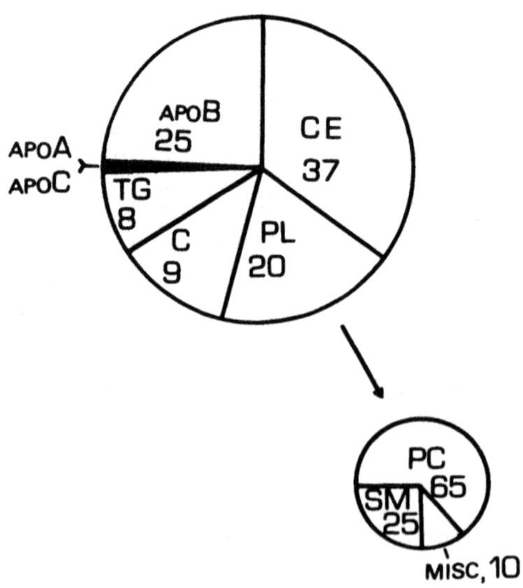

Figure 3: Lipid-protein composition of LDL.

Table 2

Average Fatty Acid Composition of LDL of Four Normal Subjects on an Ad Lib Diet

Fatty acid[a]	Phospholipids	Cholesteryl esters	Triglycerides
14:0	0.57	1.04	2.24
16:0	31.2	14.8	28.6
18:0	18.4	0.96	4.53
16:1	trace	2.53	3.87
18:1	14.6	21.4	41.7
18:2	25.5	56.6	17.6
18:3,20:3	3.64	trace	trace
20:4	7.09	trace	trace

[a] Same footnotes as Table 1.

C. Composition of VLDL

Figure 4 gives the lipid and protein composition of VLDL, the lipoprotein responsible for the transport of endogenous triglyceride (4). Approximately 10% of the particle is protein; these are apoB (37%), apoC-I (4%), apoC-II (6%) and apoC-III, and apoC-III_2 (40%). The latter differ in their respective contents of 1 or 2 moles of sialic acid. All three of the C proteins have been sequenced (10-13) and their primary structures are given in Figure 5.

The lipid composition of VLDL reflects its role as a vehicle for triglyceride transport. The triglycerides compose 59% of the particle weight (4). The remainder is made up of cholesterol (10%), cholesteryl esters (5%), and phospholipid (15%). The fatty acid composition of each lipid class of VLDL appears in Table 3. Palmitate is the major fatty acid of the VLDL phospholipids with approximately equal amounts of linoleate, oleate and stearate. The VLDL cholesteryl esters are mainly linoleate, oleate and palmitate whereas the VLDL triglycerides are largely palmitate and oleate (9).

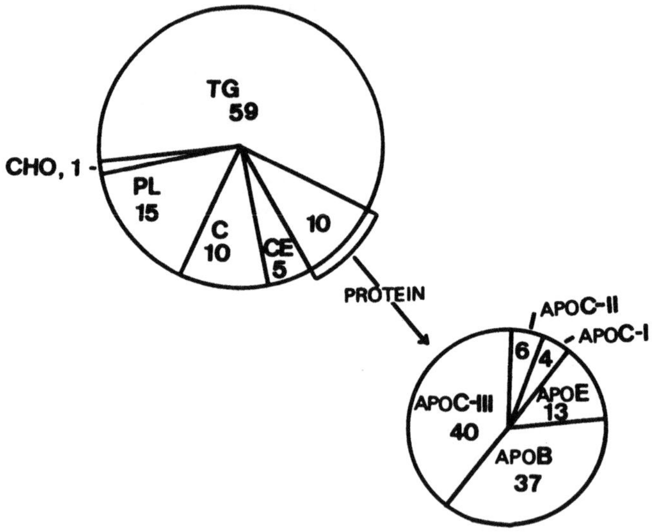

Figure 4: Lipid-protein composition of VLDL.

apoC-I	TPDVS	SALDK	LKEFG	NTLED	20
	KAREL	ISRIK	QSELS	AKMRE	40
	WFSET	FQKVK	EKLKI	DS	57
apoC-II	TEQPQ	QDEMP	SPTFL	TEUKE	20
	WLSSY	QSAKT	AAQNL	YEKTY	40
	LPAVD	EKLRD	LYSKS	TAAMS	60
	TYTGI	FTDQV	LSVLK	GEE	78
apoC-III	SEAED	ASLLS	FMQGY	MKHAT	20
	KTAKD	ALSSV	QSQQV	AAQQR	40
	GWVTD	GFSSL	KDYWS	TVKDK	60
	FSEFW	DLDPE	VRPTS	AVAA	79

Figure 5: The amino acid sequence of the C-proteins, apoC-I, apoC-II and apoC-III. Threonine-74 in apoC-III is covalently bonded to one molecule of galactosamine and galactose and 0, 1 or 2 molecules of sialic acid (designated C-III$_0$, C-III$_1$ or C-III$_2$).

Table 3

Average Fatty Acid Composition of the VLDL from Four Normal Subjects on an *Ad Lib* Diet

Fatty acid[a]	Phospholipids	Cholesteryl esters	Triglycerides
14:0	2.99	6.19	3.40
16:0	35.3	21.2	31.8
18:0	20.4	6.25	4.57
16:1	trace	3.16	5.02
18:1	17.5	23.7	41.7
18:2	20.6	35.2	13.3
18:3,20:3	trace	trace	trace
20:4	4.85	trace	trace

[a] Same as Table 1.

III. ISOLATION OF HUMAN PLASMA LIPOPROTEINS

After addition of 0.01% azide the HDL, LDL and VLDL may be isolated according to their densities using the sequential flotation technique of Scanu (15). Plasma is adjusted to a background density of 1.006 g/ml by the addition of KBr and centrifuged in polyalamar tubes using an anglehead rotor (e.g. Beckman 60 Ti) for 12 hrs. at 55,000 rpm. This floats the VLDL and, if any are present, the chylomicrons; these are removed by tube slicing. The infranatant density is readjusted to d= 1.063 g/ml and recentrifuged for 24 hrs. in a 60 Ti at 55,000 rpm. The floated LDL are again removed by tube slicing. The combined infranatant density is adjusted to d=1.21 g/ml and centrifuged for 48 hrs in a 60 Ti rotor at 55,000 rpm. The resulting supernatant contains HDL and frequently albumin. The latter may be removed by a second 48 hr. centrifugation in which the supernatant is placed in the lower half of the centrifuge tube and the upper half filled with a KBr solution of d=1.21 g/ml. The floated HDL may be removed by tube slicing and infranatant discarded.

Table 4

Properties of the Human Plasma Lipoproteins[a]

	Particle radius, nm	Density, g/ml	Density range for isolation, g/ml
Chylomicrons	35.0	0.93	1.006
VLDL	12-35	0.97	1.006
LDL_1	11-12	1.003	1.006-1.019
LDL_2	10-11.5	1.034	1.019-1.063
HDL_1	-	1.05	-
HDL_2	3.5-5	1.094	1.063-1.125
HDL_3	2-3.5	1.145	1.125-1.210
VHDL	-	1.25	1.210-1.250

[a]Adapted from Scanu et al. and ref. cited therein (2).

The lipoproteins also exist as subclasses based on their densities. These may be isolated by a similar procedure using different densities to recentrifuge a given lipoprotein class. The density ranges for the various lipoproteins are given in Table 4.

An alternative procedure proposed by Rudel et al. (15) is based on the large differences in the sizes of HDL, LDL and VLDL given in Table 4. All of the LPs are first concentrated by a 24 hr. centrifugation in d=1.225 g/ml. The concentrated LPs are then applied to a gel filtration column (Bio-Gel A5m) and are eluted in order of decreasing size, i.e. VLDL LDL HDL.

This procedure avoids the extensive centrifugation time required by sequential flotation but has the disadvantage that it has not been sufficiently refined for the isolation of the subclasses of the LPs.

IV. ISOLATION OF APOLIPOPROTEINS FROM HDL

ApoA-I and apoA-II are the major apolipoproteins of HDL and are usually derived from isolated HDL. After thoroughly dialyzing HDL against 0.1 M NaCl, 0.05 M carbonate pH 8.6 0.01% azide and 0.01% EDTA, the sample is lyophilized. The dried HDL (5g portions) are transferred to a vessels containing

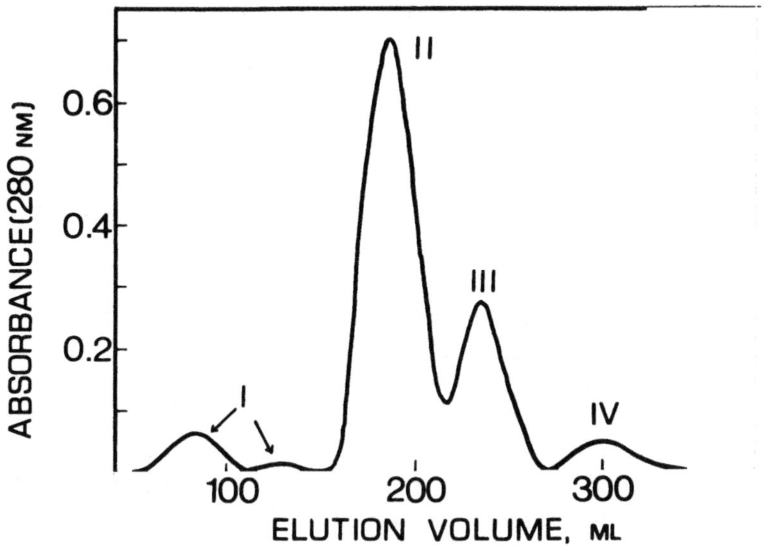

Figure 6: Chromatogram of apoHDL eluted over a 2.5 x 100 cm column of Sephadex G-200 using 8 M urea 0.1 M Tris-HCl pH 7.4 at 10^{-4} EDTA of azide.

about 100 ml of 3:1 ethanol: diethyl ether and shaken for 24 hrs. at 5°. The sample is centrifuged and the ethanol:ether is replaced daily until the protein residue shows the absence of phosphorus (16).After a final extraction with diethyl ether, centrifugation and removal of the solvent, the apoHDL is dried under a gentle stream of nitrogen.

At this point the apoHDL may be solubilized for chromatography in one of several media. These are 6.0 M guanidine HCL, (17) 8.0 M urea (18) or sodium dodecyl sulfate (19). In addition all are run in 0.1 M Tris pH 8.6 containing 0.01% EDTA and sodium azide. Up to 200 mg of apoHDL in 10 ml of the elution media may be applied to a 2.6 x 200 cm column of Sephadex G 150 eluted at a rate of 10-15 ml/hr. A typical chromatogram of apoHDL given in Fig. 6 exhibits four prominent peaks; in order of their appearance they are (I) unidentified material, (II) apoA-I, (III) apoA-II and (IV) apoC. An additional chromatography of the isolated apoA-I on apoA-II over DEAE cellulose in 6 M urea provides a highly purified protein (20). This provides ample amounts of apoA-I and apoA-II for many studies but is not the best source of C-proteins.

V. ISOLATION OF APOLIPOPROTEINS FROM VLDL

Isolation, dialysis and lyophilization of VLDL is the same as that described for HDL except that the solvent ratio for delipidation should be ethanol: diethyl ether, 1:3. This higher proportion of ether is needed to prevent solubilization of some of the C-proteins (21). After thorough delipidation and drying the suggested procedure for obtaining the individual C-proteins is the procedure of Brown et al. (22). These are as follows. (1) dissolve apoVLDL in 0.2 M Tris pH 8.2 containing 0.1 M sodium decyl sulfate. (2) elute the soluble part of this material through a Sephadex G-150 column with 0.2 M Tris pH 8.2 and 0.002 M sodium decyl sulfate. The two major peaks (Fig. 7) are firstly, B-protein and secondly A-protein and C-proteins. The C-proteins compose 60% of the total VLDL protein eluted and are freed of detergent by dialysis. The final step for isolation of the concentrated C-proteins is chromatography over DEAE in 5.4 molar urea using a salt gradient (22). A typical profile is shown in Figure 8 ; the first protein to elute from the column is apoC-I which is followed by apoC-II and finally apoC-III_1 and apoC-III_2. An alternative method of isolating C-proteins is isoelectric focusing (23).

VI. ISOLATION OF APOLIPOPROTEIN FROM LDL

ApoLDL (apoB) has proved to be very difficult to work with due to its low solubility in the absence of denaturants such as sodium docecyl sulfate and guanidine:HCl. The void volume fraction obtained from chromatography of apoVLDL in sodium decyl sulfate is relatively pure. The best source of apoB, however, is LDL in which it represents about 25% of the total lipoproteins weight. This may be isolated by solubilization in deoxycholate or by delipidation with orgainc solvents. Resolubilization is achieved only with great difficulty even if detergents are used. It is anticipated that the methodology in this are of lipoprotein purification will improve in the coming years.

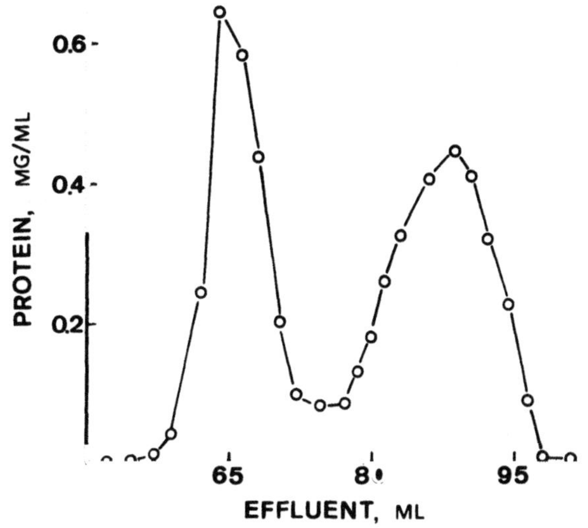

Figure 7: Chromatogram of apoVLDL eluted over Sephadex G-150 in 0.2 M Tris-HCl pH 8.2 containing 0.002 sodium decyl sulfate.

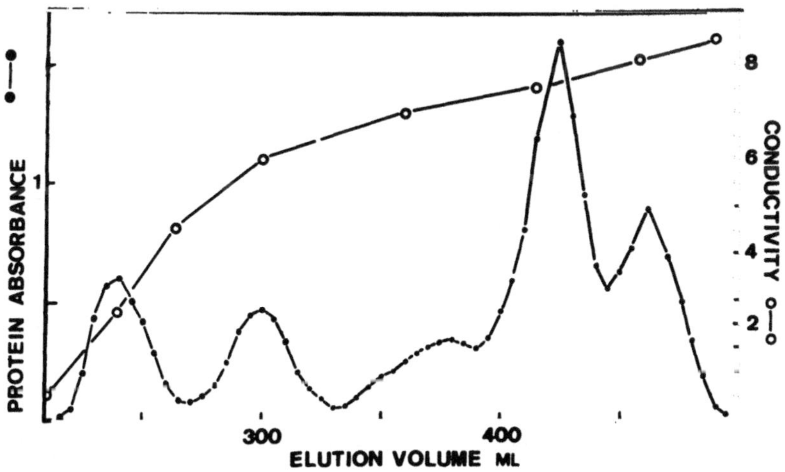

Figure 8: Representative chromatogram of apoC proteins eluted over DEAE in 5.4 molar urea using a salt gradient. The right hand cordinate indicates the salt concentration by conductivity.

VII. APOLIPOPROTEIN ANALYSIS

After pooling and concentrating the individual proteins from HDL and VLDL each pool should be checked for compositional integrity by the following methods:

1. Amino acid analysis consistent with its composition.
2. Urea or sodium dodecyl sulfate-polyacrylamide gel electrophoresis.
3. Reaction with an antibody for a given peptide and the absence of reaction with antibodies prepared from other plasma lipoproteins and albumin.

Acknowledgement

The author is an Established Investigator of the American Heart Association. A portion of this research was supported by a grant from the American Heart Association, the National Institutes of Health (HL-19459) and was developed by the Section on Atherosclerosis, Lipids and Lipoproteins of the National Heart and Blood Vessel Research and Demonstration Center, Baylor College of Medicine, a grant-supported research project of the National Heart, Lung and Blood Institute, National Institutes of Health, Grant No. HL-17269.

References

1. R.L. Jackson, J.D. Morrisett and A.M. Gotto Jr., Physiological Reviews 56 259 (1976)
2. A.M. Scanu, C. Edelstein and P. Keim in The Plasma Proteins F.W. Putnam, Ed. Academic Press New York (1975) p. 317
3. J.D. Morrisett, R.L. Jackson and A.M. Gotto Jr., Ann. Rev. Biochem. 44 183 (1975)
4. V.P. Skipski, in Blood Lipids and Lipoproteins: Quantitation, Composition and Metabolism. G.J. Nelson Ed. Wiley New York (1972) p. 471
5. S.J. Friedberg and J.A. Reynolds, J. Biol. Chem. 251 4005 (1976)
6. H.B. Brewer Jr., S.E. Lux, R. Ronan and K.M. John, Proc. Nat. Acad. Sci. USA, 69 1304 (1972)
7. H.N. Baker, A.M. Gotto Jr. and R.L. Jackson, Proc. Nat. Acad. Sci. USA, 71 3631 (1974)
8. M.O. Dayhoff, Atlas of Protein Sequence and Structure 5 Nat. Biomed. Res. Found. Washington, D.C. (1972)

9. J.D. Morrisett, H.J. Pownall, R.L. Jackson, R. Segura, A.M. Gotto Jr. and O.D. Taunton in: Chemistry and Biochemistry of Polyunsaturated Fatty Acids (W.H. Kunau and R.T. Holman, eds.) 1976
10. R.S. Shulman, P.N. Herbert, K. Wehrly and D.S. Fredrickson, J. Biol. Chem. 250 12 (1975)
11. R.L. Jackson, J.T. Sparrow, H.N. Baker, J.D. Morrisett, O.D. Taunton and A.M. Gotto Jr., J. Biol. Chem. 249 5308 (1974)
12. R.L. Jackson, H.N. Baker, E.B. Gilliam and A.M. Gotto Jr., (1977) Proc. Nat. Acad. Sci. USA, in press
13. H.B. Brewer, R. Shulman, P. Herbert, R. Ronan and K. Wehrly, J. Biol. Chem. 249 4975 (1974)
14. A.M. Scanu, J. Lipid Res. 7 285 (1966)
15. L.L. Rudel, J.A. Lee, M.D. Morris and J.M. Felts, Biochem. J. 137 89 (1974)
16. G.R. Bartlett, J. Biol. Chem. 234 466 (1959)
17. J.A. Reynolds and R.H. Simon, J. Biol. Chem. 249 3937 (1974)
18. A. Scanu, J. Toth, C. Edelstein, S. Koga and E. Stiller, Biochemistry 8 3309 (1969)
19. W.L. Stone and J.A. Reynolds, J. Biol. Chem. 250 3584 (1975)
20. B. Shore and V. Shore, Biochemistry 8 4510 (1969)
21. A.M. Scanu and C. Edelstein, Anal. Biochem. 44 176 (1971)
22. W.V. Brown, T.I. Levy and D.S. Fredrickson, J. Biol. Chem. 244 5687 (1969)
23. J.J. Albers and A.M. Scanu, Biochim. Biophys. Acta 236 29 (1971)

THE PLASMA LIPOPROTEINS IN NONHUMAN PRIMATES

V.BLATON

Simon Stevin Instituut
and
Department of Clinical Chemistry, St.Janshospitaal
B-8000 Brugge, Belgium

INTRODUCTION

The need for satisfactory animal models arises in part from the defects of man as a model for the study of atherosclerosis. Atherosclerosic lesions in humans develop irregularly over a period of years. This makes the study of either progression or regression of the disease in humans very difficult. Animal models provide the advantage that the lesions of fairly definite age can be induced in a short period of time so that factors either aggravating them or causing their regression can be studied with reasonable accuracy. The search for a completely satisfactory animal model is not finished, but an investigator now has available information concerning a large number of models from which he can choose. While no animal model exists that duplicates perfectly the human disease and all of its effects such as myocardial or cerebral infarcts, an investigator can choose from a variety of species best suited for studies of the particular facets of the disease he is interested in.

Because of their close phylogenetic relationship to man, non-human primates have become of considerable interest in a search for animal models in which metabolic and pathologic processses thought to be related to atherosclerosis simulating the situation in man. Non-human primates are a natural choice since spontaneous atherosclerotic lesions have been described in all species so far studied. Of approximately 200 species of non-human primates, 12 have been examined for the prevalence of naturally occuring lesions and in seven of these species diet induced atherosclerosis has been studied. Although numerous genera of monkeys have not been completely evaluated, those appearing to have the most potential usefulness as biomedical models in studying human atherosclerosis research are the New World Genera Saimiri, Cebus and Pan Troglodytes and the Old

World Genera Papio and Macacus. During the past decade, primary emphasis has been on Rhesus monkeys (Macaca Mulatta), Squirrel monkeys (Saimiri Sciureus), and baboons (Papio cynocephalus).

The search for the completely satisfactory animal model is not finished, but in most studies there are not enough detailed biochemical data available to distinguish one species from another. However it is opportune to review the situation at the present time particularly because we have arrived at a critical point in the future use of these animals in research. This is because of a number of restraints has been imposed by various governments on the sale and exportation of non-human primates for research. These apply especially to the rhesus monkey and the baboon. There is a real need for a summary of the results of experimental work to date and for a detailed appraisal of the experimental work that might be contemplated for the future. Perhaps the most difficult problem of all is to decide whether a monkey is essential for the proposed work or whether some more convenient animal such as the rat would be equally satisfactory. An important aim of our discussion is to try to resolve this problem of selection of a suitable experimental animal for research into atherogenesis.

ANIMALS AND DIETS

We studied in our laboratory the plasma lipoproteins of the chimpanzee, the baboon and the rhesus monkey. The animals are kept at the Zoo of Antwerp.

Then chimpanzees (white faced, Pan Troglodytes) of mixed sexes (7 males and 3 females) ± 10 years of age were given a control diet. Four animals were given a control diet and six animals were fed an atherogenic diet. These animals were being investigated as part of a long term study on the effect of high colesterol diets an atherosclerosis in the chimpanzee and had been consuming the diets for 7-8 years. The controlgroup received 10gm basic control diet per Kg body weight/day, supplemented with two apples, two bananas and one-half orange. Vegetables (400g/day), one egg half a liter of skimmed milk a day were also given. The daily caloric intake amounted to 1750Kcal/day for the control group. The control diet had a lipid content of 3gm % with 0.15gm % cholesterol and a protein content of 17gm %. Palmitic (22%), oleic (31%) and linoleic (30%) were the chief fatty acids in the control diet.

Eighteen young baboons (Papio cynocephalus), three females and 15 males, 12-18 months of age, were divided in equal groups and given either a control or an atherogenic diet. These animals were also part of a long term study of the effect of hypercholesterolemic diets on atherosclerosis in the baboon and had been consuming the diets for 1.5-3 years.

The control diet consisted of rat cake supplemented with vitamin C. Palmitic (27%) oleic (23%) and linoleic (31%) are the chief fatty acids in the control diet. The total intake of the diet amounted to approximately 25g/Kg body weight/day. Sixteen male rhesus monkkeys (Macaca mulata) were fed the same diet. (V.Blaton and H.Peeters, 1976).

BLOOD SAMPLING

It is always a difficult task to collect blood from the chimpanzee and the baboon. (Mortelmans J., 1965).
All animals were tranquilized with sernylan (Parke Davies) or with thalamonal (Janssen Pharmaceuticals) and fasting blood into EDTA (1mg/ml) was always taken from the femoral artery. Plasma was obtained after centrifugation at 3.000g and used with 6hr for lipoprotein isolation. For lipid and fatty acid analysis samples were kept at 4°C (Blaton et al, 1974).

BIOCHEMICAL DATA OF THE LIPID AND LIPOPROTEIN ANALYSIS

1. Lipid and lipoprotein analysis

In "Blood lipids and lipoproteins : quantitation, composition and metabolism", edited by Gary Nelson, we described our integrated approach to plasma lipid and lipoprotein analysis (V.Blaton and H.Peeters, 1971). Electrochromatography has been used for years in our laboratory (1965-1972) as a rapid efficient and inexpensive means of separating lipoproteins into two groups on a paper curtain. In a second stage, the fractions are cut out and eluted, their lipids are extracted and the polar and non-polar lipids classes are separated by thin layer chromatography and FA analyzed by GLC. Since elution of lipids from lipoproteins, separated by ECG with chloroform-methanol (2:1, V/V) results in the loss of 10% of phospholipids, a new extraction procedure with chloroform-methanol and acetic acid was developed. The curtain fragments containing the high and low density lipoprotein fractions from approximately 2ml of plasma are lyophilized for 15hr and then extracted with 100ml chloroform-methanol 1:1 (V/V) for 15hr with constant stirring. After a second extraction for 2 hr with the same solvent, the last traces of phospholipids are extracted with methanol-acetic acid, 99:1 (V/V). Recovery of individual lipids by elution from the paper was 99.2%. All extracts are combined and evapored under nitrogen at 40°C. Thin layer chromatography was used for the separation of polar and apolar lipids with the classical procedures. However, the apprearence of the Iatroscan using chromaroads a large number of new applications of TLC are available. In the last months the Institute developed a new quantitative and qualitative TLC method on Iatroscan for lipid fractionation (Vandamme et al, 1977).
We started years ago on a program of comparative analysis of primate plasma from animals kept at the Zoo of Antwerp under an identical diet and the same hygienic conditions. In the first experiment following experimental animals are considered chimpanzees, dwarf chimpanzee, male mountain gorilla, lowland gorillas, baboons, rhesus and other species.

A comparative study of the lipoproteins separated at that time on paper shows a progressive increase in the alpha to beta relationship. All primates have more alpha lipoprotein than man. The mountain gorilla has however lower beta Lp than the lowland gorilla and slightly more alpha lipoprotein than man. In one of the species there were two fractionated HDL lipoproteins (Blaton et al, 1968; H.Peeters et al, 1968).

Excepting the baboon all animals show high total lipids. In general the relative contribution of each lipid class is of the same order of magnitude with a few exceptions. The cholesteresters dominate the picture as could be expected from the absolute figures, but the phospholipids are percentually more important than in man, although the C/PL ratio is slightly higher than the normal human ratio.

Low lipid values are obtained for the baboon. Total cholesterol is mainly transported by LDL in the chimpanzee and in the rhesus. In the baboon however, HDL is the chief carrier of cholesterol. As for man and chimpanzee phosphatidylcholine (PC) is 72% and is the predominant phospholipid component in the baboon and the rhesus, followed by Sphingomyelin (Sphm) which is much higher in the new world species (10%). The minor phospholipid subclasses as phosphatidylethanolamine (PE), inositol (PI) and phosphatidyl serine (PS) are identical in the non-human primates and higher in concentration (mg%) than in man. Differences observed in the PL subclasses are related to structural and compositional differences in the individual apolipoproteins, which contain potential differences in functional aspects.

We characterized also the lipid composition in alfa and beta-lipoproteins, which ressemble the human pattern, except for the distribution of the phospholipid subclasses. In earlier reports no differences were described for the percentage lipid composition of the plasma Lp subclasses in primates, with a high ressemblance to the human values (Blaton et al, 1976).

2. Differentiation of fatty acids in plasma lipoproteins

The nature of the fatty acids in each lipid class is of particular interest, since the physical and chemical properties of the lipid molecules depend on the lenght of the hydrocarbons chains and the number of double bounds. Thus the metabolic properties and physiological characteristics of lipoproteins may be correlated with variations in fatty acid composition. The procedure is described by Blaton et al (1974). The fatty acid compositions of serum lipoproteins from fasting normal human subjects and control primates have been compared. The values obtained over an interval of several months indicate the constancy during this period. In comparison with chimpanzees and baboons, human subjects demonstrated a greater content of unsaturated fatty acids in both α-and β-lipoproteins.

3. Comparative properties of the major HDL apoproteins in non-human primates

The objective of the study on the comparative properties of the major HDL apoproteins in non-human primates was the isolation and partial characterization of the plasma HDL lipoproteins in order to elucidate the relationship between the apolipoproteins from man and non-human primates and to stress the usefulness of the animals in the study of metabolism, structure and function of lipoproteins.

On a modified agarose electrophoresis system, the plasma lipoproteins are separated. Baboon plasma contained VLDL and compared to the human pattern decreased low density lipoproteins. HDL are the main lipoproteins in the chimpanzee and the baboon. As a result the % concentration of the baboon total HDL (74%) and the rhesus HDL (68%) were much higher than in man (32%) and chimpanzee (53%). In comparison to man the non-human primates showed an inverse HDL_2/ HDL_3 ratio.

The purified THDL ($1.063<d<1.210$) from the non-human primates yielded a percentage protein-lipid distribution similar to the normal data in man and had a protein content of 47.5%. Surprising findings were in the percentage distribution of the phospholipid classes. As for man and as for plasma (PC) is the major PL, but the baboon and the rhesus have higher values. Differences in the non-human primates were in the higher percentage PC and the lower sphingomyelin (Spm) content for the old world representatives.

The apo-HDL peptides from human and non-human primates were fractionated on Sephadex and on DEAE-cellulose columns. As we know that baboon and rhesus are characterized by a high amount of high density lipoproteins in comparison to man and chimpanzee, it was tempting to relate the low sphingomyelin to PC-ratio in the former animals to their lipoproteins.

The major component A-I identical for all species, appears in different polymorphic forms on DEAE-cellulose using a tris gradient. ApoA-I accounts for about 60% of apoHDL in man and chimpanzee and for 70% in the baboon. In all four species considered, apoA-I is very similar both qualitatively and quantitatively. The second polypeptide, apoA-II, represents 20% of apoHDL in man and about 12% in chimpanzee. In addition the chimpanzee apoA-I and apoA-II reacted with monospecific antisera raised in the rabbit against either of the two human products. In the baboon and the rhesus the apoA-II is eluted at a different volume from the DEAE column which means that it represents some difference with the two other species. Actually the main difference between man and baboon apoA-II appears to be that the human peptide is present in the dimeric form whereas in the baboon it is a monomer. The similarity between chimpanzee apoA-I and man also extended to the microhetogeneity of apoA-I which was documented by PAGE-isoelectrofocusing.

The amino acid composition of apoA-I in the four species is identical (no cysteine, no Il except glutamine and methionine) and shows a complete resemblence to the major HDL apoprotein between both species.

The amino acid composition of apoA-II in man and chimpanzee is identical. The difference between the dimeric apoA-II in man and chimpanzee and the monomeric apoA-II in the baboon and the rhesus is the presence of arginine and the absence of halfcystine in the baboon and the rhesus apoA-II which is substituted by serine suggesting that the disulfide bridge is not required for the lipid binding functions of the protein.

The apoproteins A-I of the four species were treated with carboxy-peptidase A indicated glutamine, the amino acid most rapidly released as the COOH-terminal residue. Threonine and serine were also rapidly liberated but complete release after 30 minutes of digestion was not achieved. Leucine was released at a slower rate. The kinetics of release of the amino acids suggest a COOH-terminal sequence of LEU-Ser-Thr-Gln for a man and the non-human primates. For the NH_2 terminal sequence analysis, the apoA-I proteins were subjected to 30 cycles of the automatic Edman degradation. Human and chimpanzee have the same sequence. The baboon and the rhesus have identical sequences with the exception of residues 15 and 21. In the human apoA-I there is alanine at residue 15 and valine at residue 21.

The amino acid sequence of HDL apoA-II in man is compared to that of baboon and rhesus. The chimpanzee shows a AA sequence identical to man, while the baboon and rhesus have a 77 residue polypeptide which exhibits a great degree of homology with the monomeric form of human A-II. The major difference between the dimeric apoA-II and the monomeric form is the presence of Arg and the absence of half cystine, which is substituted by serine. Additional human to monkey substitutions are Lys 3 → Glu 3, Seri 40 → Ala 40, Ile 53 → Val 53, Gly 71 → Asp 71, most of which are conservative except for position 3 where glutamic acid replaces lysine. The substitutions occuring in the rhesus and baboon were found to have a higher helical potential than the corresponding residues in the human form. This is especially significant for the alanine predicted helical regions, consisting of residues 25-30 and 33-49 appear to be identical in both the human and rhesus proteins.

The structure of native baboon apoA-I is highly ordered with a helical content of 67% compared to 55% in human apoA-I, which is identical for the rhesus monkey.

ApoA-I at a concentration of 36 M is a monomeric form, above 36 M two distinct components with $S°_{20W}$=3.9, are observed which show an identical AA composition when separated by gel permeation chromatography, suggesting self association of apoA-I. The ultracentrifugation behaviour of rhesus A-I differs markedly from that reported for human apoA-I, where apoA-I self associates in solution according to a monomer dimer tetramer-octamer model, at significantly lower protein concentration than those observed for rhesus apoA-I. In our laboratory, studies on the baboon apoA-I show rather monomer dimer self association as in the human and the affinity for phospholipids is higher for the dimeric apoA-I. The marked difference in the mode of self association between human and rhesus apoA-I is also reflected by the capacity of each of these apoproteins to bind lipids in vitro. Studies from Scanu's laboratory have shown that under comparable experimental conditions of protein and lipid concentrations, rhesus apoA-I binds significantly more lipids than does human apoA-I, this observation has been related to the degree of self association of these apoproteins in vitro. It is thus apparent that, as we have pointed out before, the solution properties of apoA-I and other apolipoproteins must be known before they can be used for meaningful ligand-binding studies.

In order to investigate whether monomeric baboon apoA-II presented the tendency to aggregate reported for human apoA-II, the concentration dependence of the mean residue ellipticity was followed over a 100-fold concentration range. The negative mean residue ellipticity at 222nm of monomeric baboon apoA-II is concentration dependent and increases when the monomer to dimer equilibrium is shifted towards higher molecular weight species by increasing the concentration. The maximum value (θ222 = -10.240) agrees with that measured for the human apoprotein and corresponds to an α-helical content of 37%. Upon titration with increasing quantities of DMPC vesicles the mean residue at a ratio of 50mol of DMPC/mol of apoA-II. The maximal value corresponds to an α-helical content of 54%. This figure higher than that reported for the titration of human apoA-II with egg PC, suggesting that the structure of baboon apoA-II is stabilized by phospholipid to a larger extent than that of the human apoprotein. The cysteine residue in human apoA-II seems not to be involved in self association nor in phospholipid binding as its substitution with serine in the baboon does not significantly affect the behaviour of the apoprotein.

The enthalpy changes which evolved on binding DMPC liposomes to human and baboon apoA-I are plotted as a function of the phospholipid/protein molar ratio. DMPC binding to baboon apoA-I is an exothermal process at any phospholipid to protein ratio. The maximal enthalpy change is slightly lower for the baboon than for the human apoprotein while the complex composition is very similar for both species. The enthalpy changes which evolved on binding DMPC to baboon apoA I are exothermal at low lipid/protein ratios, in contrast with the behaviour of human apoA-I. The enthalpy titration as resulting human apoA-I DMPC was interpreted as resulting from the superposition of an endothermal process predominant at high PL/apoprotein molar ratios. The sigmoidal shape of the curve for baboon apoA-I indicates that these two effects are also present but that the magnitude of the endothermal process is less than for the human apoprotein. This endothermal process was attributed to a lipid induced disaggregation of apoA-I in agreement with other reports.

The enthalpy changes which evolved on binding DMPC liposomes to native baboon monomeric and human dimeric apoA-II are studied. The complex formed between DMPC and either human or baboon apoA-II has a maximal composition of 35 mol of DPMC/mol of human monomeric apoA-II and 24 mol for the baboon which are quite similar.

The heat changes observed on binding shingomyelin liposomes at 41° to human and baboon apoA-I are very low 20Kcal/mole at PL/prot ratio of 100M/mole. This might result from the small amount of apoprotein incorporated in the complex which was observed by gradient ultracentrifugation. For apoA-II the enthalpy changes compare with those obtained with DPMC. The results of these experiments indicate that the PL binding to human and baboon apoproteins is very similar. The complex composition as well as the enthalpy changes involved in complex formation are of the same order of magnitude. However, the hypothesis remain open for the low SPM content in the baboon (Blaton et al, 1974, 1977; Rosseneu et al, 1977).

CONCLUSION

Living primates, a group of mammels which constitute a series of successively more advanced organisms have proved to be very useful in the study of structure and functions of the plasma lipoproteins. The studies however, reported thus far on the serum lipoprotein distribution of various non-human primates are not sufficient to permit any definite conclusion. Only recently attention has been directed to the study of the serum lipoprotein apoproteins. The most detailed studies up to now have been conducted on the HDL apoproteins of Macacus rhesus (Edelstein et al, 1972; Scanu et al, 1972). Data of baboon and chimpanzee HDL apoproteins were described and for detailed results we refer to the described references. These lipoprotein particles were found to have properties similar to those of man.

REFERENCES

1. V.BLATON, H.PEETERS
The non-human primates as models for studying atherosclerosis. Studies on the chimpanzee, the baboon and the rhesus maccacus. Atherosclerosis Drug Discovery Book, 67, 33 (1976).

2. J.MORTELMANS
Primates Med., 2, 113 (1969).

3. V.BLATON, D.VANDAMME, M.VASTESAEGER, J.MORTELMANS, H.PEETERS
Dietary induced hyperbetalipoproteinemia in chimpanzees : Comparison to the human hyperlipoproteinemia, Experimental and Molecular Pathology, 20, 132 (1974).

4. V.BLATON, H.PEETERS
Integrated approach to plasma lipid and lipoprotein analysis, Chapter 6 : in "Blood Lipids and Lipoproteins", G.Nelson, ed., 369-431 (1971) J.Wiley and Sons, New York.

5. D.VANDAMME, V.BLATON, H.PEETERS
Screening of plasma lipids by thin-layer chromatography with flame ionization detection on chromarods, Journal of Chromatography, 145 (1978).

6. H.PEETERS, V.BLATON
Lipid fatty acid relationship in electrochromatographic lipoprotein fractions,
Prog. Biochem. Pharmacol., 4, 144-152 (1968), Karger Basel.

7. V.BLATON, H.PEETERS, G.A.GRESHAM, A.N.HOWARD
Differential fatty acid composition of alpha and beta lipoproteins in baboons
Prog. Biochem. Pharmacol., 4, 122-125 (1968), Karger Basel.

8. V.BLATON, R.VERCAEMST, N.VANDECASTEELE, H.CASTER, H.PEETERS
Isolation and partial characterisation of chimpanzee plasma hich density lipoproteins and their apolipoproteins,
Biochemistry, 13, 1127 (1974).

9. V.BLATON, R.VERCAEMST, M.ROSSENEU, J.MORTELMANS, R.L.JACKSON, A.M.GOTTO Jr., H.PEETERS
Characterization of baboon plasma high density lipoproteins and of their major apoproteins,
Biochemistry, 16, 2157 (1977).

10. M.ROSSENEU, V.BLATON, R.VERCAEMST, F.SOETEWEY, H.PEETERS
Phospholipid binding and self-association of the major apoprotein of human and baboon high density lipoproteins,
Eur. J. Biochem., 74, 83-88 (1977).

ACKNOWLEDGEMENT

The author wishes to thank Dr.H.Peeters, Director of the Simon Stevin Instituut and Dr.Sc.M.Rosseneu for the helpful discussion.

This research was supported by grants of NFGWO (Brussels, Belgium, grant nr 1026 and nr 30.0175).

I am grateful to Mrs.N.Vinaimont, Mr.D.Vandamme, Mr.B.Declercq and Mr.R.Vercaemst for their excellent technical assistance.

Structure of the Lipoprotein Molecule

PHYSICAL PROPERTIES, CHEMICAL COMPOSITION AND STRUCTURE OF CIRCULATING LIPOPROTEINS

Ferenc J. Kézdy

Department of Biochemistry
University of Chicago
Chicago, Illinois 60637

Before discussing the structure of lipoproteins, a brief survey of the surface properties of their major constituents is in order. This will be followed by a discussion of the basic thermodynamic laws governing small spherical particles. The main discussion will be concerned with the gross structural features of lipoproteins, which can be predicted from the regularities observed in chemical composition as a function of particle size.

I. SURFACE PROPERTIES OF PHOSPHOLIPIDS AND APOPROTEINS

A. Phospholipids

The pronounced amphiphyllic character of phopholipids with moderate or long aliphatic chains (lecithins, sphingomyelins, etc., with more than eight carbons in the side chains), allows them to form unusually stable mono-molecular layers at the air-water interface (1). The molecular packing and the pressure dependency of the compressibility of these monolayers depends on the degree of unsaturation of the aliphatic chains. Saturated chains, such as the ones in palmitic or stearic acid, are conducive to tightly packed, quasi-solid monolayers, where the molecules are aligned on the surface with their side chains in a hexagonal packing. The presence of double bonds in the side chain, especially the _cis_ bonds of oleic and linoleic acid, results in less well packed layers, where the fluidity of the monolayer is maintained even at very high surface pressures. Typically, unsaturated phospholipids occupy a limiting area of the order of 65-70 Å/molecule under surface pressures of 30-40 dynes/cm. Even under these conditions, the monolayer is liquid, i.e. the chain-

chain interactions are limited to random hydrophobic forces similar to the ones observed in liquid hydrocarbons.

Surface potential measurements also indicate the absence of well-defined phase transitions over a wide limit of surface pressures for unsaturated phospholipid monolayers. It is this property of being able to form layers which remain liquid over a large range of conditions, which renders unsaturated phospholipids eminently suitable as basic constituents of biological surfaces. Liquid phospholipid monolayers are readily miscible with other monolayer-forming molecules, such as cholesterol and proteins. At low surface pressures (2 to 10 dynes/cm) the addition of cholesterol exerts a "condensing effect" i.e. the average area occupied in the mixture by each molecule is decreased by some 10 to 20% (2). The condensing effect is much less pronounced at high surface pressures and the molecules tend to occupy their respective limiting areas, about 40 $Å^2$/molecule for cholesterol.

B. Apolipoproteins

At the air-water interface, the surface properties of apolipoproteins are rather unique in comparison with other proteins. Globular proteins, such as chymotrypsin lysozyme, or subtilisin, form quasi-irreversibly denatured monolayers with low collapse pressures (of the order of 5 dynes/cm). The formation of these monolayers on the surface of the protein solution is slow and presumably the denaturation of the protein is rate limiting. In contrast, apolipoproteins apo-A-I and apo-A-II form monolayers with a purely diffusion-controlled rate from dilute solutions ($10^{-7}$to 10^{-8}M), the formation of the monolayer is reversible, and the limiting pressures are of the order of 20 to 25 dynes/cm (3). The partitioning of the protein between the solution and the surface obeys a thermodynamic equilibrium. For each concentration in solution corresponds a given surface concentration, and the equilibrium can be obtained either by starting with the protein in solution only or in the surface monolayer only. At surface pressures of less than 2 dynes/cm the reversibly formed protein monolayer is unstable and it slowly denatures, until the protein reaches very large limiting areas. At higher surface pressures, apo-A-I and apo-A-II occupy areas corresponding to 15 $Å^2$/residue. Finally, the apoproteins are able to reach the air-water interface even when the surface is covered by a phospholipid monolayer of moderate pressure. The penetration of the protein into the phospholipid layer is indicated by a rapid increase in surface pressure, until a limiting pressure of 20 to 25 dynes/cm is reached (3,4).

II. THERMODYNAMIC RELATIONSHIPS (5)

Circulating lipoproteins are spherical particles with diameters ranging from 10^{-7} to 10^{-8} cm. The small size and the large curvature of the surface result in far-reaching modifications of the thermodynamic properties, as compared with large particles of the same chemical composition. For a liquid phase containing a solute Gibbs-Duhem equation states that

$$-\Gamma = (C/RT)(\partial \gamma / \partial C)$$

where Γ is the surface excess of the solute in moles/cm^2, γ is the surface tension of the solution in dynes/cm and C is the molarity of the solute. Thus the concentration of the solute at the surface is different from that in the bulk phase and this difference is reflected in the concentration dependency of the surface tension of the solution. Experimental observation of the latter indicates that for aqueous solutions the surface excess is negative for salts, since the surface tension increases with increasing salt concentrations. For organic molecules in general the surface excess is positive; their local concentration close to the surface is larger than in the bulk phase. By measurment of the optical anisotropy of the solutions in the surface region it is possible to show that the thickness of the "Gibbs layer" is somewhere between 20 and 100 A, depending somewhat on the solute. Thus, the anisotropy due to the surface is transmitted to molecules several molecular diameters away from the surface and these asymmetric forces profoundly influence the chemical and physical properties of all molecules close to the surface. Since for lipoproteins the majority of the components are located in the Gibbs layer, the extrapolation of properties observed in bulk solutions to lipoproteins must be accomplished with due caution.

Lipoproteins are spherical particles indicating that a positive interfacial tension between the water and the particle decreases the surface until a minimal surface corresponding to an energy minimum is reached. At equilibrium, the compressive forces of the surface tension are compensated by an internal pressure. This pressure inside the particle can be calculated from the Young-Laplace equation:

$$\Delta P = 2\gamma/r$$

where ΔP is the difference of pressures inside the particle and in the solution and r is the radius of the spherical particle. Application of this equation to lipoproteins shows that if the interfacial pressure is 20 dynes/cm for HDL the internal pressure is of the order of 100 atm.

A further consequence of the surface curvature is the modification of the chemical potential of the molecules in a small sphere. The Kelvin equation states, that

$$\mu - \mu_f = 2\gamma V/r$$

where μ is the chemical potential of the liquid in a spherical particle, μ_f is that of the same compound in a large bulk phase, and V is the molar volume of the compound. Thus, a decrease in particle size results in an increase of the chemical potential, i.e. the partial molar free energy. Since the chemical potential governs all chemical equilibria and many of the physical properties, the chemical reactivity and the transferability of a molecule in a small particle is much larger than in the bulk phase. As an extension of the above relationship, the equation derived by V.K. La Mer and R. Gruen (6) states that

$$\ln X_r = \ln X_f - (2\gamma V/RT)(1/r)$$

where X_r is the mole fraction of a solute in a spherical particle of radius r and X_f is that of the same solute in the bulk phase (6). This equation shows that when a solute is distributed between particles of different size, the concentration of the solute decreases with decreasing radii.

In summary, the large curvature and the small size of lipoproteins confers special properties to lipoprotein constituants, properties not directly measurable using the same components in a bulk phase. Most importantly, the thermodynamic relationships point to the important role of the lipoprotein surface as a determinant of the physico-chemical properties of the constituant molecules.

III. CHEMICAL COMPOSITION AND STRUCTURE OF LIPOPROTEINS

The chemical composition of circulating lipoproteins show a considerable variation in the percentage of the major components

Table I. CHEMICAL COMPOSITION BY DRY WEIGHT PERCENTS OF HUMAN SERUM LIPOPROTEINS

Particle	$MWX10^{-6}$	Protein	Phospholipids	Triglycerides	Cholesteryl ester	Free cholesterol
$VHDL_1$	0.15	62	28	4.6	3	0.3
HDL_3	0.18	55	23	4.1	12	2.9
HDL_2	0.36	41	30	4.5	16	5.4
LDL	2.3	21	22	11	37	8
VLDL	19.6	8	18	50	12	7
CHYL	504	2	7	84	5	2

for the different density classes of lipoproteins. As shown in Table I, (7) increasing particle size is accompanied by a monotonous decrease in protein content, an increase in triglyceride, and a decrease in phospholipids for the larger particles. On the other hand, the cholesteryl ester and free cholesterol contents pass through a maximum for LDL.

These regularities in the composition suggest that there is a casual relationship between particle size and chemical composition, i.e. for a given particle size there is just one chemical composition which satisfies the requirements imposed by the structure. If this is indeed the case, then the structural features can be deduced from the compositional data. For such an undertaking, the data in Table I were first converted into numbers of molecules of each constituent per particle, using the estimated molecular weights of 850 for triglycerides, 775 for phospholipids, 650 for cholesteryl esters and 387 for cholesterol (Table II). Since proteins in the different lipoproteins have widely different molecular weights, the percent of protein was expressed in terms of the number of amino acids per particle, using an average residue weight of 100. In addition to the data presented in Tables I and II the numerical analysis also included some twenty other lipoprotein compositions reported for subclasses of VLDL and lipoproteins of a variety of vertebrates (8).

The starting point of the structural analysis was the determination of space filling requirements of the hydrophobic constituents i.e. triglycerides and cholesteryl esters. If these constituents occupy a hyrophobic spherical core at the center of the lipoprotein, then the sum of their volumes should be linearly related to the cube of the radius of the particle (r) decreased by the thickness of an amphiphillic surface layer (x). Thus the number triglyceride mole-

Table II. MOLECULES OF COMPONENTS PER PARTICLE

Particle	Amino acid	Phospho-lipid	Free cholesterol	Cholesteryl ester	Trigly-ceride	Equivalent radius (A)
$VHDL_1$	961	56	1.2	8	8.3	37.5
HDL_3	963	51	13	32	9.5	39.2
HDL_2	1476	137	50	90	19	50.8
LDL	4830	653	475	1310	298	95.9
VLDL	15656	4545	3539	3600	11500	200
CHYL	102000	45160	25840	27700	507000	600

cules per particle (n_{tg}) and that of cholesterol esters (n_c) should obey the equation:

$$n_{tg}v_{tg} + n_{ce}v_{ce} = (4/3)\pi(r-x)^3 \quad (1)$$

where v_{tg} and v_{ce} are the molecular volumes of triglycerides and cholesteryl esters respectively. Since the ratio v_{ce}/v_{tg} should be equal to the ratio of the molar volumes of the two species of molecules (= 0.68) Equ. 1 can be transformed into the linear form:

$$(n_{tg} + 0.68\ n_{ce})^{1/3} = (4\pi/3v_{tg})^{1/3}(r-x). \quad (2)$$

The data for all lipoproteins indeed yielded an excellent straight line when analyzed according to Equ. 2 (correlation coefficient = .997) and the following parameters were obtained: v_{tg}=1556 $Å^3$/molecule; v_{ce}=1068 $Å^3$/molecule and x=20.5 Å. Thus for all lipoproteins, the triglycerides and cholesteryl esters occupy a spherical core of diameter r-20.5 Å and the packing of these hydrophobic molecules corresponds to that of the pure constituents in the liquid state.

Preliminary analysis indicated that the surface of the hydrophobic core is too small to accomodate the proteins as well as the amphiphillic phospholipids and cholesterol. On the other hand the surface of the core is correlated with the number of phospholipid (n_{pl}) and cholesterol molecules (n_c) alone. The following equation was tested:

$$s_{pl}n_{pl} + s_c n_c = 4\pi(r-x)^2 \quad (3)$$

where s_{pl} and s_c are the molecular areas of phospholipids and cholesterol respectively. Taking for the ratio s_c/s_{pl} the value observed for their ratio in a tightly packed monolayer at the air-water interface (=40.5 Å /71 Å =o.57), Equ. 3 can be transformed into a linear relationship:

$$(n_{pl} + 0.57n_c)^{1/2} = (4\pi/s_{pl})^{1/2}(r-x). \quad (4)$$

Numerical analysis again indicated that Equ. 4 is an excellent description of the phospholipid and cholesterol composition of all lipoproteins, with the following numerical parameters: s_{pl}=68.5 $Å^2$/molecule; s_c=39.1 $Å^2$/molecule and x=20.2 Å. Thus only phospholipid and cholesterol are in contact with the hydrophobic core, and these surface molecules are packed on the inner surface to the same extent as that found in their highly compressed monolayers at the air-water interface.

In the light of the preceding considerations, the outer surface of the lipoprotein particle in contact with water should be occupied then by proteins as well as by phospholipids and cholesterol. Preliminary analysis indicated however, that the area occupied by cholesterol on the outer surface is negligibly small, i.e. free cholesterol is unable to reach the lipoprotein-water interface. Under these conditions the surface-filling requirements impose the following relationship:

$$s'_{aa}n_{aa} + s'_{pl}n_{pl} = 4\pi r^2 \qquad (5)$$

where s'_{aa} and s'_{pl} are the area occupied by an average amino acid residue and phospholipid molecule respectively.

Equ. 5 can be transformed into the linear form:

$$r^2/n_{aa} = s'_{aa}/4\pi + (s'_{pl}/4\pi)(n_{pl}/n_{aa}). \qquad (6)$$

The experimental compositional data indicated excellent correlation with Eq. 6 and the following parameters were obtained: s'_{aa}=15.6 Å^2/residue and s'_{pl}=62.7 Å^2/molecule. Thus, the lipoprotein-water interface is occupied exclusively by phospholipids and proteins, and cholesterol must be covered by polypeptide chains.

Finally, the cholesterol content of all lipoproteins is in excellent agreement with the La Mer-Gruen equation. Indeed, numerical analysis of the cholesterol content of the surface layer of lipoproteins indicates that cholesterol is a true solute in this layer and that the interfacial tension between water and lipoproteins is 39 dynes/cm. It should be pointed out that the data for chylomicrons deviate from the theoretical equation, suggesting that the rapid degradation of chylomicrons prevents their equilibration with the other lipoproteins in the circulating plasma.

The preceding analysis demonstrates, that the chemical composition of all lipoproteins is in excellent agreement with the space- and surface-filling requirements of a structural model composed of a hydrophobic core and amphiphillic surface monolayer. There never was much doubt about the triglycerides and cholesteryl esters occupying a centrally located core, but the present model stresses their clear segregation from the rest of the lipoprotein components. Also, the analysis reveals the existence of a well defined surface monolayer, the components of which do not interpenetrate with triglycerides and cholesteryl esters. The numerical results obtained for the thickness of this monolayer (20 Å) and the various molecular area are in excellent agreement with those observed for monolayers of the same components at the air-water interface. These eminently reasonable numerical results further support the validity of the proposed model.

The essential features of the proposed model are in agreement with the large majority of experimental observations concerning the physical and chemical properties of lipoproteins. On the other hand, a space- and surface-filling analysis is inherently unable to make predictions concerning the chemical and physiological role of the variety of subclasses of components, such as the sphingomyelins, lecithins and phosphatidyl ethanolamines, or the different apoproteins. It is most likely, however, that the structural role of these subclasses is very similar, if not identical.

ACKNOWLEDGEMENTS

The work presented in this article was generously supported by Grants 1 P17 HL 15062 and 1 P01 HL 18577 from the National Heart and Lung Institute, U.S. Public Health Service. The author wishes to thank Dr. Betty W. Shen and Dr. Angelo M. Scanu for their contributions.

REFERENCES

1.) For a review, see e.g. G.L. Gaines, Jr.: "Insoluble Monolayers at Liquid-Gas Interfaces". Interscience, N.Y., 1966.
2.) D. Chapman, N.F. Owens, M.C. Phillips and D.A. Walker; Biochim. Biophys. Acta 183, 458 (1969).
3.) B.W. Shen, A.M. Scanu and F.J. Kézdy; Circulation, 7, Suppl. IV 218 (1973).
4.) M.C. Phillips, M.T.A. Evans and H. Hauser; Adv. in Chemistry Series 144, 217 (1975).
5.) For a review, see e.g. A.W. Adamson: "Physical Chemistry of Surfaces". 3rd Edition, Interscience, N.Y., 1976.
6.) V.K. La Mer and R. Gruen; Trans. Farad. Soc. 48, 410 (1952).
7.) From A.M. Scanu and A.W. Kruski: "International Encyclopedia of Pharmacology and Therepeutics". Pergamon, Oxford Vol. 1, sect. 24, 21 (1975).
8.) B.W. Shen, A.M. Scanu and F.J. Kézdy; Proc. Natl. Acad. Sci. USA 74, 837 (1977).

THE PHYSICO-CHEMICAL PROPERTIES OF LIPIDS IN LIPOPROTEINS

M.C. Phillips

Unilever Research Laboratory, The Frythe, Welwyn, Herts., U.K.

I. INTRODUCTION

As can be seen from Table 1, serum lipoproteins comprise a variety of lipid molecules. The density of the particles decreases as the content of neutral lipid increases. Thus, low density lipoproteins (LDL) and chylomicrons contain a preponderance of triglycerides and cholesterol esters which are immiscible with water. Apart from chylomicrons, the various lipoprotein classes contain roughly the same proportion of polar, water-dispersible lipids (phospholipids). Clearly, any physico-chemical discussion of the properties of lipoprotein lipids must consider their mixing behaviour, as well as the properties of the pure, anhydrous species and their interactions with water.

Table 1

The Principal Components of Human Serum Lipoproteins[a]

Component[b] (% w/w)	HDL_3	LDL	VLDL	Chyl
Protein	~ 56	21	10	2
Cholesterol	3	8	10	3
Cholesterol Esters	12	37	5	2
Glycerides	4	11	~ 55	~ 88
Phosphatidylcholine	17	14	~ 12	3
Sphingomyelin	3	6	~ 5	2
Non-esterified fatty acids	2			

a. Adapted from data in ref. 1,2
b. Components present at $\lesssim 1\%$ level are not included

This paper first considers the physical states which hydrocarbon chains can adopt, dealing in particular with polymorphism and the effects of temperature. After this, the consequences of the hydrophobicity of hydrocarbon chains for the phase behaviour and surface-activity of phospholipids in water is dealt with. Finally, the miscibility of polar and neutral lipids is discussed; particular emphasis is paid to the effects of altering the ratio of phosphatidylcholine to cholesterol on the emulsification of neutral lipids in water.

II. THE BEHAVIOUR OF HYDROCARBON CHAINS

The C–C single bonds in a free alkane chain can take up three stable conformations of which the two gauche conformations are higher in energy than the trans conformation (for reviews, see ref. 3-5). In a condensed lipid phase, intra- and intermolecular steric restrictions limit the number of possible chain conformations. At low temperatures the chains are likely to be in the low energy all-trans conformation (Fig. 1). As the temperature is raised, there will be some probability of gauche rotations about C–C bonds with energies typically about 0.5 kcal mol^{-1} appearing in the chains. In a molecule such as butane, where the energy barrier preventing rotation between trans and gauche conformations is ca. 3.6 kcal mol^{-1}, the jump frequency

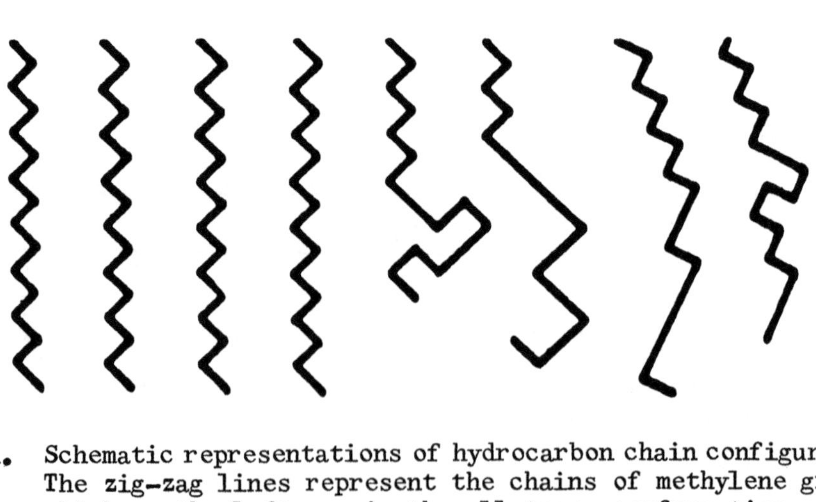

Fig. 1. Schematic representations of hydrocarbon chain configurations. The zig-zag lines represent the chains of methylene groups which on the left are in the all-trans conformation. The remaining chains have several gauche rotations in the lower half and show different angles of tilt.

between the two states is ca. 10^{10} s^{-1}. Steric interactions between chains prevent most single bond trans-gauche changes until relatively high temperatures when a cooperative melting of the chains occurs.

Table 2

Thermodynamic Data for the Thermal Transitions of Compounds Containing n-Hexadecane Chains

Compound	Ordered Form*	Melted Form	T_c °C	ΔH kcal mol^{-1}	ΔS cal mol^{-1} deg^{-1}	Ref.
n-hexadecane	β crystal	isotropic liquid	18	13	44	27
palmitic acid	β "	"	63	13	39	58
	β "	"	66	43	127	
tripalmitin	β' "	"	45	30	94	58
cholesteryl palmitate	crystal	cholesteric mesophase	78	15	42	31
	cholesteric mesophase	isotropic liquid	80	0.3	1	31
dipalmitoyl phosphatidylcholine						
- monohydrate	β' crystal	smectic mesophase	65	5	14	27
- in excess water	β' gel	"	41	9	28	27

*The nomenclature of Luzzati[14,46] and coworkers is employed here.

Table 2 contains thermodynamic data describing the chain melting of a series of lipids containing n-hexadecane chains. Since all the compounds melt at higher temperatures (T_c) than n-hexadecane it is clear that the presence of polar groups stabilises the crystal phases of these lipids. The T_c increases with increasing chain length and the more unsaturated the hydrocarbon chains the lower is the T_c; cis double bonds have much greater effects than trans double bonds[6]. The predominant fatty acids in serum lipoproteins are palmitic (n-hexadecanoic acid), oleic (cis-9-octadecenoic acid) and linoleic (cis-9-,cis-12-octadecadienoic acid)[2]; the melting points of these species are 63 (β crystal-Table 2), 16 (β crystal) and -5°C, respectively.

Besides the contribution from the increase in internal energy of the hydrocarbon chains due to rotational isomerism, the rise in enthalpy (ΔH) at the T_c contains a contribution from the change in intermolecular cohesive and repulsive forces arising from expansion of the lipid (the $p\Delta V$ work term is negligible). It is apparent from Table 2 that ΔH per hydrocarbon chain for the transition from β crystal to isotropic liquid (melt) is constant at about 13 kcal mol^{-1}; thus ΔH is proportional to the number of single C–C bonds involved in the transition. The sharpness of the phase transition is of interest because it depends on the number of molecules forced to cooperate in the transition; the latter number is called the cooperative unit and for a first order transition is infinity. However, in real crystals imperfections can cause local melting so that the cooperative unit is smaller than the total solid. The Van't Hoff enthalpy (ΔH_{VH}) relates to the reversible, two-state transition of a cooperative unit. It is possible to derive ΔH_{VH} from differential scanning calorimetry traces of excess specific heat through the temperature range of the transition because

$$\Delta H_{VH} \approx 6.9 \frac{T_m^2}{\Delta T_{\frac{1}{2}}} \qquad [1]$$

where T_m = mid-point temperature and $\Delta T_{\frac{1}{2}}$ = half-height width in degrees of the specific heat curve[7]. Integration of the area under the excess specific heat curve gives the calorimetric enthalpy of transition (ΔH_{cal}); this corresponds to the transition enthalpy per single molecule. Thus, the ratio of ΔH_{VH} to ΔH_{cal} gives a convenient measure of the cooperativity of the transition and may be interpreted as the number of lipid molecules in the average cooperative unit. Using this approach, the cooperative unit in synthetic phosphatidylcholine bilayers when dispersed in water as liposomes is ~ 100 molecules[8]; this is not necessarily the overall size of a local patch of solid or fluid molecules[9]. Since the cooperative unit in lipoproteins cannot be larger than a single particle and these frequently contain less than 100 lipid molecules of a given type[2], any chain-melting transitions induced in lipoproteins are likely to be of low cooperativity and therefore broad relative to the transitions in liposomes containing a single type of phospholipid molecule. This means that there will be a temperature range in which crystalline and melted lipid molecules coexist in lipoprotein particles.

The data for tripalmitin in Table 2 show that different chain-melting transitions can be observed at 45 and 66°C. This arises because lipids such as tripalmitin can adopt different crystal forms i.e. they exhibit polymorphism.

Polymorphism

Lipids generally crystallise so that the molecules are arranged in layers bounded by planes of end-groups (terminal methyl groups, polar head groups). The molecular arrangement is the result of the interplay of the packing requirements of the polar groups and the long hydrocarbon chains. These effects can lead to a variety of crystal structures and polymorphism is common with lipids (for reviews, see ref. 10-13). When one of two polymorphs is thermodynamically unstable at all temperatures below T_c, the system is said to be monotropic. Only the transformation from metastable to stable polymorph can be observed directly and the metastable polymorph can often be obtained only by transformation from metastable (undercooled) liquid. When each of two polymorphs is thermodynamically stable in a definite range of temperature and pressure they are said to be enantiotropic; a transition between the two polymorphs occurs at a definite temperature. It should be noted that the distinction between monotropic and enantiotropic polymorphism can often be blurred by kinetic effects.

The results of X-ray powder studies of lipid crystals have shown that n-alkane chains adopt the trans configuration and various angles of tilt with respect to the plane of the end groups,

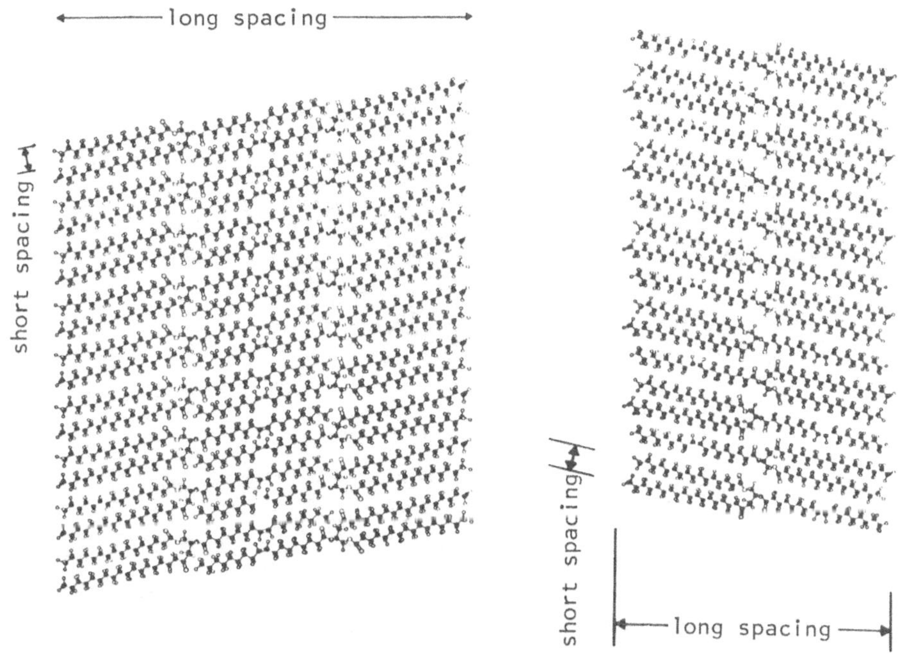

Fig. 2. The **right** diagram illustrates the double chain packing of the triglyceride (POS) containing palmitic, oleic and stearic acid chains in the 1-,2- and 3-positions, respectively. The left diagram represents triple chain packing of POS. Clearly, the X-ray long spacings are quite different in the two cases.

but the conformation of the polar groups cannot be obtained by this technique. Several nomenclatures have been used to describe the various polymorphs and we shall follow Luzzati[12] in the subsequent brief description of the structures adopted by hydrocarbon chains. In the β form the hydrocarbon chains are stiff and fully extended and organised with rotational disorder according to a two-dimensional hexagonal lattice; the β' conformation is similar except that the chains are tilted with respect to the normal to the plane of the lamellae (cf. Fig. 1; see ref. 12,14). The exact crystal packing often depends upon the solvent used for recrystallisation. Examples of some polymorphic forms of triglycerides are depicted in Fig. 2 (cf. 11). The chain conformation above the melting point is designated α.

Knowledge of the detailed conformation of the whole lipid molecules in crystals requires the complete elucidation of the structure of a single crystal. By way of example, we describe briefly the results of an X-ray diffraction analysis by Hitchcock et al.[15,16] of 1,2 dilauroyl-DL-phosphatidylethanolamine crystals grown from glacial acetic acid. The monoclinic crystals contained one molecule of acetic acid per lipid molecule and were of mediocre quality by ordinary crystallographic standards in that the long range order was poor. The one-dimensional electron density pro-

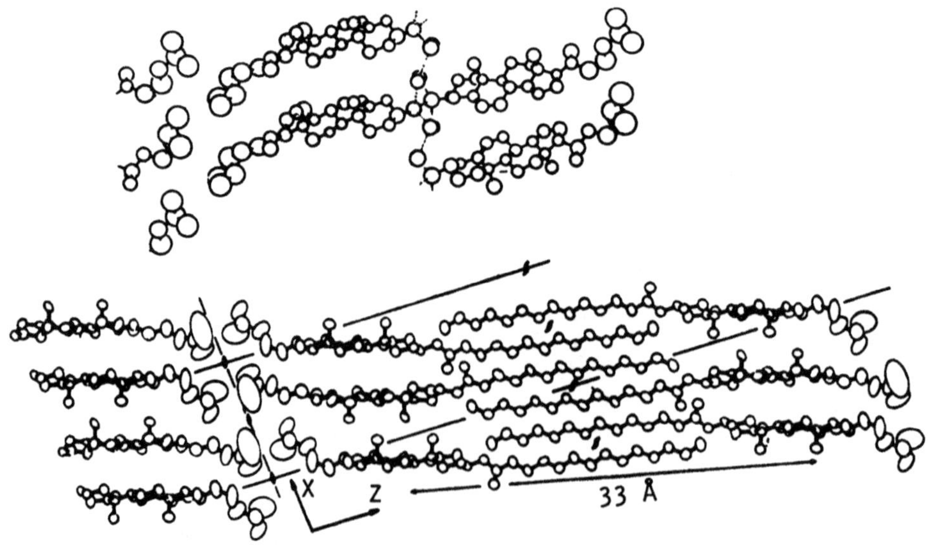

Fig. 3. Top: crystal structure of cholesterol monohydrate (from ref. 19). Bottom: crystal structure of cholesteryl myristate (from ref. 20). The atoms have a 50% probability of being enclosed by the circles or ellipsoids.

files are characteristic of those from phospholipid bilayer membranes (cf. 17,18). Thus, the intermolecular packing of the phosphatidylethanolamine molecules produces the classical lipid bilayer structure, adjacent bilayers being separated by acetic acid molecules of crystallisation. The lipid chain axes are essentially parallel to one another and there is a progressive increase in disorder towards the centre of the bilayer. Within one section of bilayer, the separation of the centres of gravity of the polar groups is 39.0Å. The P-N axis of the zwitterion is approximately parallel to the plane of the bilayer and there is an intermolecular hydrogen bond of length 2.79Å between the primary amino group and the phosphate group.

Recent X-ray single crystal studies by Craven and coworkers[19,20] have yielded the crystal packings of cholesterol monohydrate and cholesterol myristate (see Fig. 3). The structure of cholesterol monohydrate shows a stacking of bilayers which are 33.9Å thick. The ring systems of all cholesterol molecules are nearly parallel and the C-3 hydroxyl groups and water molecules are hydrogen-bonded to form a pleated sheet at the bilayer interface. The C-17 chain atoms have large thermal mean square amplitudes (Fig. 3) and a reversible endothermic ($\Delta H = 0.7$ kcal mol^{-1}) transition occurs between 35 and 40°C which is related to a change in the packing of these terminal CH_3 groups[21]. As shown in Fig. 3, these groups are also highly mobile in cholesterol myristate crystals. In this case, the structure consists of stacked bilayers 50.7Å thick in which cholesterol myristate molecules are packed in an antiparallel fashion so that the cholesterol residues are adjacent with the myristate chains interdigitated. The latter chains are all trans and so in the β configuration.

Since the hydrocarbon chain packing is the same in the β′ forms of tripalmitin and dipalmitoyl phosphatidylcholine, the fact that in the latter case ΔH and ΔS for chain melting are lower (Table 2) shows that the melted form of phosphatidylcholine (i.e. the smectic mesophase) is more ordered than liquid tripalmitin. The reasons for this are discussed in the next section.

Thermotropic Mesomorphism

Many compounds do not pass directly from the crystal to the liquid state on heating, but exhibit one or more intermediate liquid-crystalline or mesomorphic states[22-24]. Three types of mesomorphic states are generally recognized and these are depicted in Fig. 4. Smectic liquid crystals are made of sheets of orientated molecules; fluidity arises because the sheets are flexible and can slide over one another. Nematic liquid crystals are not as highly ordered as smectic mesophases; the molecules are arranged with their long axes parallel but not separated into layers. The molecules apparently can be drawn past one another in the direction of their long axes, a degree of freedom which is excluded in smectic phases by the comparatively strong attractions between the ends of the molecules. As the name implies, the cholesteric mesophase is exhibited by many cholesterol derivatives; the molecules in each layer are parallel but because of the shape of cholesterol molecules, the orientation of the molecules in each layer is displaced progressively. The overall displacement traces out a helical path and this has an effect on light transmitted perpendicular to the molecular layers so that circular dichroism occurs.

Only phospholipids and cholesterol esters of the lipids which occur in serum lipoproteins (Table 1) form mesophases; both exhibit smectic mesophases and cholesterol esters also form cholesteric liquid crystals. The thermotropic mesomorphism of phospholipids has been investigated in great detail (for reviews, see 3,6,12,13, 18,25). The lamellar phase is comprised of bimolecular leaflets of hydrocarbon chains covered on both sides by the polar groups.

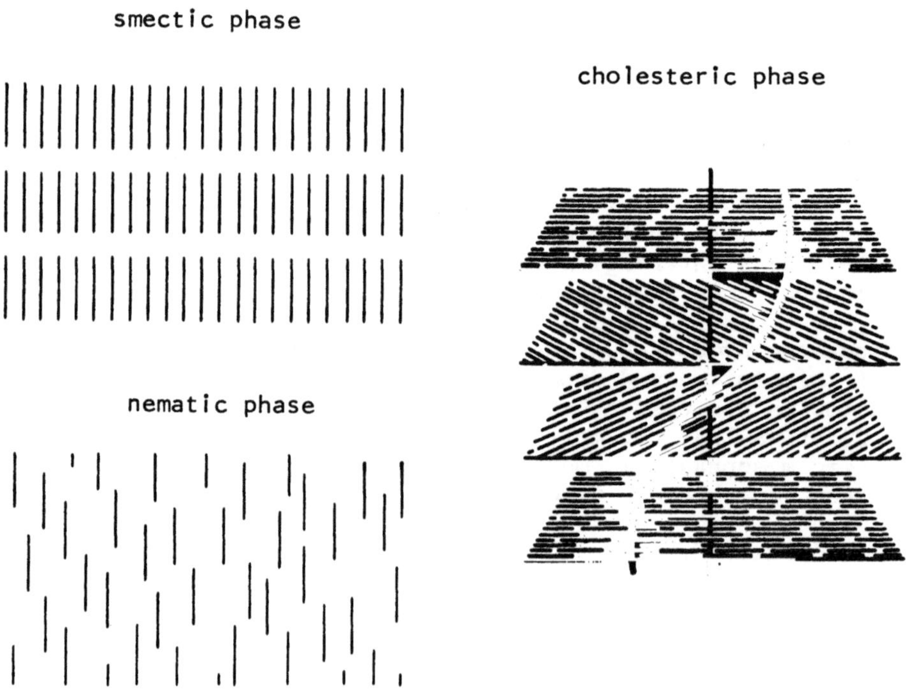

Fig. 4. Schematic representations of the molecular arrangements characteristic of the various forms of thermotropic liquid crystal. The helicity of the overall displacement of molecular orientation from layer to layer in the cholesteric mesophase is indicated.

The lamellae are parallel and equidistant. In the hexagonal phase the polar groups form cylinders of indefinite length which are packed in a two-dimensional hexagonal lattice, the hydrocarbon chains occupying the region between the cylinders. In the cubic phase, identical rodlike elements are linked three by three into two three-dimensional networks which are unconnected but mutually interwoven. The physical state of the hydrocarbon chains in the bilayers of the predominant smectic phase is of particular interest. As can be seen from Table 2, ΔH and ΔS per hydrocarbon chain for the transition of dipalmitoyl phosphatidylcholine from the β' crystal to smectic mesophase are approximately half the equivalent figures for chain-melting of tripalmitin where the isotropic liquid is formed. This reduction in the configurational freedom of the hydrocarbon chains in anhydrous phospholipid mesophases, as compared to the fluidity of liquid n-alkanes at the transition temperature, is a consequence of the order imposed by the polar group lattice[3,26,27]. The chains are anchored at one end to the most rigid part of the bilayer, the glycerol backbone of the phosphatidylcholine molecules, and in this region the possibility of trans-gauche isomerisations is reduced because the steric restriction is greatest in this region[28]; the configurational freedom and mobility of the chains increases towards the centre of the bilayer[10] so that the terminal methyl groups have correlation times of ~ 10^{-10} s which is similar to that in liquid n-alkanes[29].

Cholesterol does not form any mesophases and melts directly to an isotropic liquid at 147°C. In accord with the observation[30] that ΔH and ΔS for the melting of rigid molecules is generally lower than for flexible molecules because of the lower possible configurational disorder, the ΔH of 6.5 kcal mol^{-1} and ΔS of 15 cal mol^{-1} deg^{-1} for melting cholesterol[31] are lower than the equivalent figures for lipids such as n-hexadecane or palmitic acid (Table 2). Cholesterol esters behave differently in that they form mesophases (for reviews, see 23,31,32); some of the thermodynamic data for cholesteryl palmitate are presented in Table 2. A cholesteric mesophase is formed on heating the crystals to 78°C and isotropic liquid forms at 80°C. Cooling the isotropic liquid leads to the reversible formation of cholesteric mesophase at 78°C which converts to smectic mesophase at about 74°C (ref. 31); this phase is monotropic and gradually converts to the crystal phase. As with the melting of simple fatty acids, the transition temperatures of cholesteryl esters are a function of the degree of unsaturation of the fatty acid moieties. For example, Davis and coworkers[33] have found that for the stearate, oleate, linoleate and linolenate series, the crystal to isotropic liquid transition temperatures are 82,50,42 and 35°C, respectively.

The structures of the liquid crystals formed by cholesteryl esters are not known although it has been shown that some characteristics of the crystal structure are retained[20]. The data for cholesteryl palmitate in Table 2 suggest that formation of the mesophases is accompanied by some trans-gauche isomerisation in the palmitate chain (cf. 34). The packing of the cholesterol residues in cholesterol monohydrate and cholesteryl palmitate crystals is similar (Fig. 4) so that ΔH and ΔS for disordering these regions are likely to be similar in both cases. Thus, to a first approximation, the differences in ΔH and ΔS for melting of cholesterol and cholesteryl palmitate can be attributed to the palmitate chain which is in the β configuration in crystals of the latter compound. This approach suggests that the ΔH and ΔS values associated with disordering the palmitate chain in cholesteryl palmitate are ~ 8 kcal mol^{-1} and 27 cal mol^{-1} deg^{-1}, respectively (this applies to formation of cholesteric and smectic mesophases, or isotropic liquid, because ΔH and ΔS for the transition between these three states are relatively small[31]). The above values are ~ 60% of the equivalent figures for the melting of β crystals of n-hexadecane or palmitic acid suggesting that the palmitate chains in the cholesteric mesophase are more ordered than in isotropic liquid of these simple lipids. Overall, the data in Table 2 suggest that the average fluidity of the palmitate chain in the cholesteric mesophase of cholesteryl palmitate is intermediate to that of the fully hydrated smectic mesophase of dipalmitoyl phosphatidylcholine and the isotropic liquid of palmitic acid. Direct spectral observations of chain motions in cholesterol ester mesophases are required to check this conclusion.

III. INTERACTIONS OF LIPIDS WITH WATER

It is well known that "oil and water don't mix" and the "hydrophobic effect" which give rise to this demixing has been the subject of much study (for reviews, see 35,36). The free energy rise associated with mixing a hydrocarbon chain with water is a consequence of complex changes in water structure which are not well understood. Recently, it has been established[37,38] that the free energy due to hydrophobic interactions for the homologous series of n-alkanes is proportional to the area of apolar surface exposed to the aqueous phase. The hydrophobic interfacial free energy (i.e. the free energy associated with each unit area of hydrocarbon surface in contact with water) is 20-33 cal mol^{-1} $Å^{-2}$. Because of this interfacial free energy, the emulsion formed by dispersing droplets of liquid n-alkane in water is highly unstable and phase separation occurs rapidly. Pure triglyceride and cholesteryl ester oils are similarly immiscible with water because the ester bonds are not sufficiently polar to outweigh the hydrophobic effect.

Increasing the polarity of a lipid causes it to be either water-dispersible of water-soluble. Such molecules are amphiphilic because they have a polar "head" which mixes with water and a hydrocarbon chain "tail" which is immiscible with water. Because of this dual solubility characteristic, amphiphilic molecules adsorb strongly to interfaces (Fig. 5) (i.e. they are surface-active). Surfactant molecules can be classified with respect to their hydrophilic-lipophilic balance on an "H.L.B." scale. Such values, which increase with the polarity of the molecule, represent an empirical numerical correlation of the emulsifying and solubilising properties of different surfactants[39].

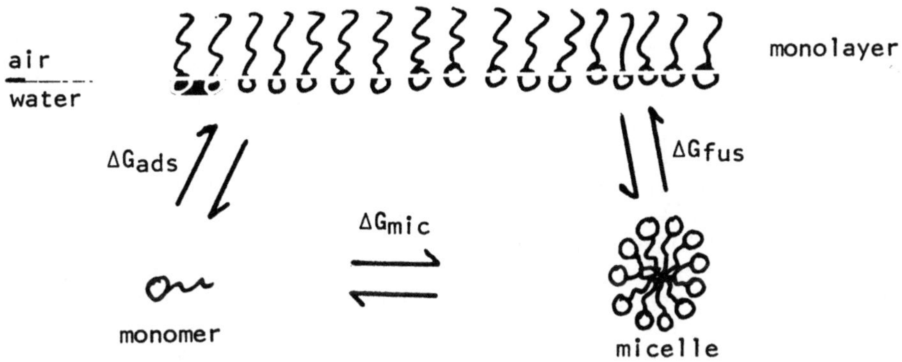

Fig. 5. Schematic representations of the possible locations of hydrophobic lipid molecules when added to water. The possible equilibria are shown.

Fig. 5 summarises the behaviour of lipid molecules which are sufficiently polar to dissolve in water to give significant concentrations of monomer. Lipids such as soaps have to be heated above a critical temperature called the Krafft point before they dissolve. The hydrophobicity of the hydrocarbon chains causes adsorption at the air-water interface and formation of micelles. The hydrophobic free energy is proportional to the number of CH_2 groups so the free energies of adsorption (ΔG_{ads}) and micellisation (ΔG_{mic}) vary linearly with the length of hydrocarbon chain. The chain-length dependences of these two functions and the free energy of transfer (ΔG_{trans}) of surfactant from water to n-alkane oil are given in cal mol^{-1} by the following equations[35].

$$\Delta G_{trans} = k_1 - 850\, n \qquad [2]$$

$$\Delta G_{ads} = k_2 - 625\, n \qquad [3]$$

$$\Delta G_{mic} = k_3 - 700\, n \qquad [4]$$

n is the number of carbon atoms in the hydrocarbon chain and the constants are a function of the polar group. These equations are qualitatively consistent and indicate that the free energy of transfer of a CH_2 group out of contact with water is about - 700 cal mol^{-1} in each of the three processes considered; the differences are due to slight variations in the chain interactions when they are in monolayers, micelles and n-alkane oils.

When amphiphilic molecules are dissolved in water they can achieve segregation of their hydrophobic portions from the solvent by self-aggregation to form the micelles depicted in Fig. 5 (ref. 35). Inverted micelles with a polar group core form when the surfactant is dissolved in a non-polar oil. In aqueous systems the formation of micelles is opposed by a repulsive force originating from the polar groups; the repulsion can involve electrostatic and/or hydration effects. These repulsions oppose the attractive force arising from the hydrophobic effect thereby limiting the size to which micelles can grow; for simple amphiphiles containing a single hydrocarbon chain the aggregation number is of the order of 100. As the total concentration of amphiphile is increased, a transition from the predominantly unassociated amphiphile to the micellar state occurs over a narrow critical range of concentration which is called the critical micelle concentration ("cmc"). Raising the concentration of amphiphile above the cmc leads to an increase in the proportion of amphiphile in micelles while the concentration present as monomer remains more or less constant (if the temperature is below the Krafft point, solid surfactant precipitates above the cmc). When the monomer and micelles are in equilibrium, there is rapid exchange of molecules between the two states.

Some effects of variations in the hydrocarbon chains and polar groups are demonstrated by the cmc's listed in Table 3. For an homologous series such as the sodium soaps, log cmc is a linear function of n; the cmc decreases by a factor of ~ 0.3 when n is increased by two[40]. The importance of polar group interactions is demonstrated by the fact that neutral palmitoyl lysophosphatidylcholine forms micelles more easily than the n-hexadecyl sulphate and trimethylammonium molecules which carry a net charge. The much lower cmc for dipalmitoyl phosphatidylcholine as compared to the lyso compound is consistent with the above variation of cmc with n (or a hydrophobic free energy of - 700 cal mol^{-1} CH_2 group); this shows that the very low cmc for dipalmitoyl phosphatidylcholine arises from the presence of two chains[41]. In this case, the aggregation number is ~ 3000 and the structures of these micelles

Table 3

The Effect of Variations in Chain Length and Polar Group on the cmc's of Lipids in Water

	Temp°C	cmc (M)	Ref.
Sodium soap homologues			
decanoate	50	100×10^{-3}	59
dodecanoate	50	24×10^{-3}	59
tetradecanoate	50	7.3×10^{-3}	59
hexadecanoate	50	2.1×10^{-3}	59
n-hexadecyl (C_{16}) surfactants			
trimethylammonium bromide	25	80×10^{-5}	59
sodium 1-sulphate	25	20×10^{-5}	59
lysophosphatidylcholine	25	1×10^{-5}	60
di C_{16} phosphatidylcholine	20	5×10^{-10}	41
cholesterol	25	3×10^{-8}	42

or liposomes have been the subject of much study (see below). Cholesterol, which has a maximum solubility in water of ~ 5×10^{-6}M, starts to form rod-like micelles of aggregation number ~ 500 at the exceptionally low cmc of ~ 3×10^{-8}M. The free energy of stabilisation of the micelles contains a contribution due to strong intermolecular attractive forces as well as the hydrophobic effect[42].

A simple surfactant such as sodium palmitate needs a free concentration of ~ 10^{-3}M to form an adsorbed monolayer (which is soluble), saturate the surface and give the maximum reduction in surface tension; this effect is due to the adsorption of monomer which at the cmc and above is ~ 10^{-3}M. Dipalmitoyl phosphatidylcholine forms an insoluble monolayer which can exist in a variety of phases[3] so, unlike sodium palmitate, the equilibrium between the surface and the substrate is established relatively slowly. This metastability of the monolayer and slow rate of equilibration is presumably associated with the very low concentration of monomer (~ 10^{-10}M) present in this system and low probability of collision with the surface. The fact that at relatively high concentrations of ~ 1% w/w phosphatidylcholine (~ 10^{-2}M), the surface tension of the air-water interface is reduced significantly suggests that molecules may transfer directly from the micelles (liposomes) to the interface by some fusion process (Fig. 5). This needs further study but transfer of molecules between membranes and liposomes by this process is known to occur.

When water is added to relatively hydrophobic lipids such as phospholipids they swell and disperse in water rather than dissolve as the monomer. Such behaviour is called lyotropic mesomorphism in that the particular phase obtained is function of both temperature and water content. The formation of these structures is a phenomenon which involves both melting and solution[43] and the polarity of the lipid plays a critical role[44]. There is a minimum temperature below which water cannot penetrate into the crystal lattice of polar lipids[45]. At this penetration temperature (T_c, Krafft point), the hydrocarbon chains melt and water diffuses into the polar region of the crystal lattice. Once this occurs the phospholipid crystal can swell in water, and myelin tube formation can proceed spontaneously[45]. The T_c depends upon the nature of the hydrocarbon chains and of the polar region of the molecule, the amount of water present, and on any solutes present in the water. The chain-melting phase transition which occurs in anhydrous lipids also occurs in the hydrated systems so that when the sample is cooled to below T_c, the hydrocarbon chains arrange themselves into an orderly crystalline lattice, but the water is not necessarily expelled from the system. Lyotropic phases containing crystalline paraffin chain regions are called gels (see Fig. 6a); these gels may or may not be metastable. If they are metastable they transform over a period of time to a suspension of microcrystals of the amphiphile in water - the coagel state.

Many phase diagrams, based upon X-ray diffraction experiments carried out with lipid-water samples of different composition as a function of temperature, are now available in the literature (for reviews, see 12,13,17,18,25). Utilising the notation of Luzzati and coworkers[46], the one-dimensional lamellar phases (Fig. 6a,b) are designated L, while the two-dimensional hexagonal structures (Fig. 6c,d) are specified H. In hexagonal I, the cylinders of phospholipid molecules with the hydrocarbon chains to the interior are separated from each other by water, whereas in the inverse hexagonal II phase the cylinders are embedded in a matrix of hydrocarbon chains. Two-dimensional oblique or rectangular lattices are designated P. The homogeneous conformations of the hydrocarbon chains specified by α, β, β' have been described earlier. A mixture of α and β hydrocarbon chain conformations is designed $\alpha\beta$ when the relative content varies with temperature and/or concentration; a phase in which the relative content of the two types of conformation is fixed by the symmetry of the lattice is called γ (ref. 47).

For all the lipids considered in Fig. 7, the bilayer is the predominant structure when more than about 20 wt% water is present. The progressive increase in the lamellar spacings of dimyristoyl and egg phosphatidylcholines with increasing water finally ceases with the formation of two phases when about 45 wt% water is present. The lateral expansion of the bilayer due to the incor-

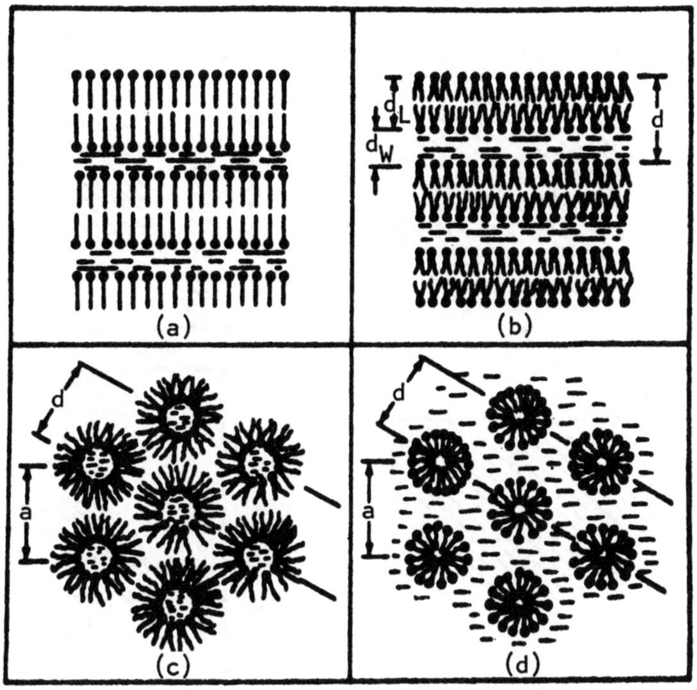

Fig. 6. Schematic representations of lipid-water phases. (a) gel; (b) lamellar liquid-crystalline; (c) hexagonal type II; (d) hexagonal type I.

poration of water[48] leads to thinning of the bilayer. The phase behaviour of sphingomyelin-water systems[49,50] parallels that of phosphatidylcholine-water mixtures; in both cases limited amounts of water are incorporated into the polar group regions, which carry no net charge at physiological pH's, of the bimolecular lamellae. The lamellar phase also predominates in phosphatidylethanolamine-water systems with a hexagonal phase (H II) appearing only when the temperature is elevated[18]. Unlike phosphatidylcholine, phosphatidylethanolamine is not hygroscopic and does not swell as strongly when mixed with water; phosphatidylethanolamine is fully hydrated above T_c when ~ 20 wt% water is added, whereas excess water only appears with phosphatidylcholine when more than 40 wt% water is present. This increased accommodation of water can be attributed in part to the greater lateral expansion of phosphatidylcholine bilayers which arises from differences in the orientation of the zwitterions in phosphatidylcholine and phosphatidylethanolamine[51]. The net charge on acidic phospholipids leads to them having dramatically different swelling properties to a

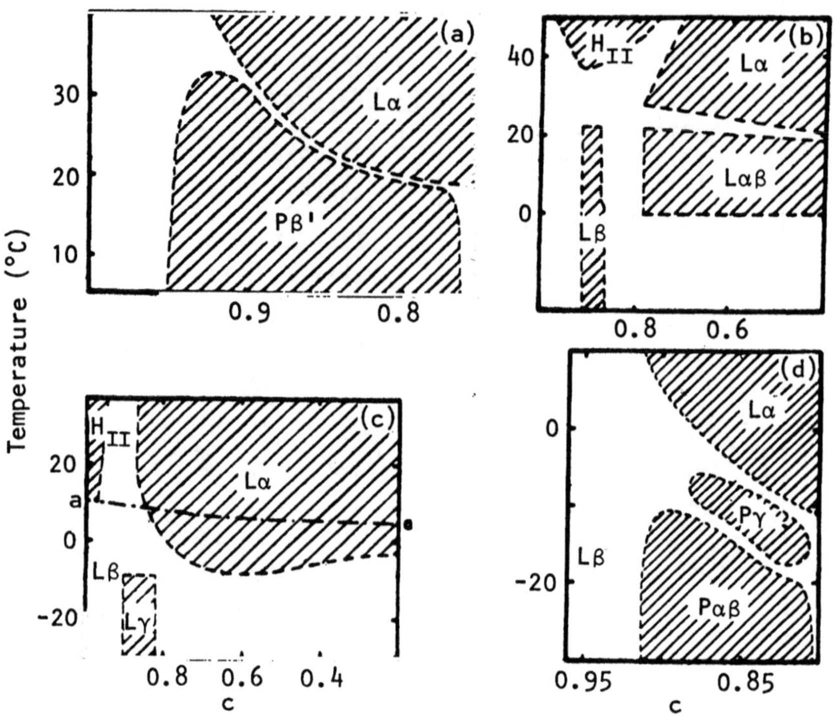

Fig. 7. Portions of the phase diagrams of various lipid-water systems. The regions where one phase is found pure are hatched. c is the weight concentration:lipid/(lipid + water). (a) Dimyristoyl phosphatidylcholine; (b) egg phosphatidic acid; (c) mitochondrial lipids; (d) egg phosphatidylcholine (from Ranck et al.[47]).

neutral phospholipid such as phosphatidylcholine. However, as with phosphatidylcholine, the lamellar phase is predominant; for example, egg phosphatidic acid forms bilayers exclusively when the water content is above about 25 wt% (Fig. 7b). Acidic phospholipid bilayers separate much more than bilayers comprising neutral phosphatidylcholine molecules. Gulik-Kryziwicki et al.,[52] have shown that incorporation of as little as about 1% of either cationic or anionic lipid into the lamellar phase in excess water of egg phosphatidylcholine makes the swelling behaviour comparable to that of phosphatidylserine. This enhanced swelling arises because of electrical double layer repulsion between adjacent bilayers[51]. For this reason, total phospholipid extracts from mammalian cell membranes swell to incorporate large amounts of water between the bilayers; for example, beef heart mitochondrial lipids can incorporate up to 80 wt% water (see Fig. 7c).

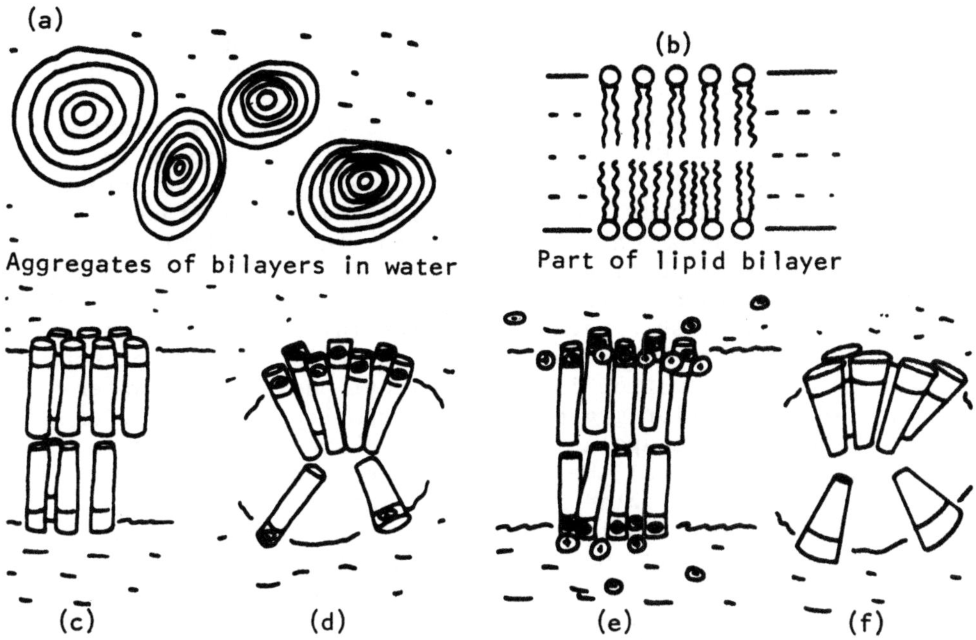

Fig. 8. Effect of shape and charge on aggregation. (a) Phospholipid aggregates present in an aqueous dispersion of phosphatidylcholine. Each concentric line in structure a represents a bimolecular leaflet, as shown in structure b. (c) Bimolecular leaflet formed by cylindrical phospholipid molecules. (d) Structure with increased curvature of the lipid-water interface formed in solution of low ionic strength by phospholipid molecules with net negative charge on the head groups. (e) Curvature of interface of structure d reduced by shielding in the presence of excess ions. (f) Spherical micelle formed by wedge-shaped molecules with or without charge. (From Phillips[3]).

The above brief outline of the phase behaviour of lipid-water systems indicates that the form of the hydrated polymolecular structures depends upon the shape and charge of the lipid molecule. The phosphatidylcholine molecule is roughly cylindrically shaped, allowing easy side-by-side stacking for attainment of the maximum interaction energy. There is no net charge on the polar head group which can give rise to a disrupting, repulsive, electrostatic energy either normal or parallel to the plane of a bilayer and as a result phosphatidylcholines form stable multilamellar aggregates of continuous bimolecular leaflets (Fig. 8a-c). Because they are insufficiently hydrophobic, phosphatidylcholines with less than 10 carbon atoms in their hydrocarbon chains will not form stable bimolecular lamellae at ordinary temperatures[3]. If the lipid molecules are ionized, a mutual repulsion between the polar ends of the molecules is introduced, and there is a tendency for the bilayer to break up into aggregates of radially oriented molecules (Fig. 8d). When phosphatidylserine is dispersed in aqueous solutions of low ionic strength, small bilayer structures are formed with high curvature of the lipid-water interface, whereas, in 0.1M sodium chloride solutions, larger multilamellar aggregates are stable because of shielding of the negative charge on the phosphatidylserine molecule by counterions (Fig. 8e). This shielding leads to less curved lipid-water interfaces and less repulsion between the bilayers. As yet there is no evidence for spherical micelle formation in acidic diacyl phospholipid-water systems. When the polar group of the phospholipid has a different cross section from its apolar hydrocarbon chains (i.e., the molecule is wedge-shaped) the maximum interaction energy is obtained with a curved polar group-water interface. When the polar group cross section is large (e.g. lysolecithin), either a hexagonal I phase (Fig. 6d) or spherical micelle (Fig. 8f) is formed depending upon the water concentration. Reduction in the size of the polar group (e.g., phosphatidylethanolamine) can result in a hexagonal II phase (Fig. 6c).

IV. MIXING BEHAVIOUR OF LIPOPROTEIN LIPIDS

The mixing properties of the lipids listed in Table 1 are clearly relevant to lipoprotein structure and we address ourselves to this topic here. Since triglycerides and cholesteryl esters do not hydrate and are immiscible with water whereas phospholipids hydrate strongly and swell in water, it is convenient to consider mixing in terms of these two categories.

Generally, the physical state of the hydrocarbon chains is crucial in determining miscibility. In anhydrous systems which are totally liquid, the lipids are completely miscible but cooling to cause a change of phase of one component can lead to some immiscibility. For example, cholesteryl linolenate can be incorporated into anhydrous egg phosphatidylcholine mesophases up to only 25% w/w; the cholesteryl ester molecules are situated in the hydrocarbon chain regions of the liquid crystal lattice and decrease the T_c of phosphatidylcholine[53]. Similarly, a feature of the interaction between cholesteryl esters or cholesterol with triglyceride is that lower melting components depress the melting point of the high melting ones[54]. Furthermore, triglyceride can remove the mesophases in mixtures with cholesterol and cholesteryl linoleate; e.g., 1-3% triolein removes the cholesteric mesophase and with $>50\%$ no mesophase is observed and crystallisation takes place directly from the isotropic liquid. The low mutual solid solubility of cholesterol, cholesteryl ester and triglyceride means that crystals of pure cholesterol separate readily at ordinary temperatures[54]. In order for lipids of a given class to form solid solutions, their hydrocarbon chains must be of similar length and degree of unsaturation so that their intermolecular interactions and melting points are close.

Addition of water can have a dramatic effect on the mixing of lipids such as cholesteryl esters and phosphatidylcholine. Janiak et al.,[55] have investigated the egg phosphatidylcholine/cholesteryl linolenate system and found that addition of water converts phosphatidylcholine into an Lα phase in which cholesteryl linolenate is soluble to only ~ 2% (i.e. 1 molecule cholesteryl linolenate to 40 phosphatidylcholine molecules). Furthermore, hydration of the phosphatidylcholine polar groups prevents these molecules entering the hydrophobic, oily cholesteryl ester phases; phosphatidylcholine separates as the hydrated Lα phase. Cholesterol can dissolve in the latter phase at levels up to 33% (i.e. up to equimolar amounts) and the behaviour of cholesterol/cholesteryl ester/phospholipid mixtures in excess water is summarised in Fig. 9. The data in Table 1 indicate that the lipid compositions of serum lipoproteins fall in Zone III and at equilibrium would form two phases, viz. phospholipid bilayer containing up to 33% cholesterol and 2% cholesteryl ester and a liquid or liquid-crystalline cholesteryl ester phase (liquid triglyceride would dissolve in this phase). The fact that LDL lipids do not separate in this way in vivo means that either the existence of LDL particles[34] involves non-equilibrium effects or that the apoproteins perturb the phase behaviour summarised in Fig. 9.

The solubilisation of the water-immiscible cholesteryl ester/triglyceride phase into water involves coating the surface of droplets of this material with hydrated polar lipid, cholesterol and protein. Thus, the formation of lipoprotein particles involves preparing a microemulsion of triglyceride/cholesteryl ester in water. Unless heated[56], such a microemulsion is stable in that coalescence of the particles and separation of a bulk oil phase does not occur. Since phospholipids can emulsify triglyceride oils into water very effectively, the presence of apoprotein as additional emulsifier seems superfluous, at first glance. However, as shown in Table 4, the cholesterol acts in an antagonistic fashion to the phosphatidylcholine and can destabilise triglyceride oil-in-water emulsions. In order to understand these effects it is necessary to consider some emulsion science.

The fact that water-dispersible phosphatidylcholine molecules stabilise oil-in-water (O/W) emulsions whereas the relatively water-insoluble cholesterol molecules stabilise water-in-oil (W/O) emulsions is consistent with Bancroft's Rule[57] which states that the phase in which the emulsifying agent is more soluble will be

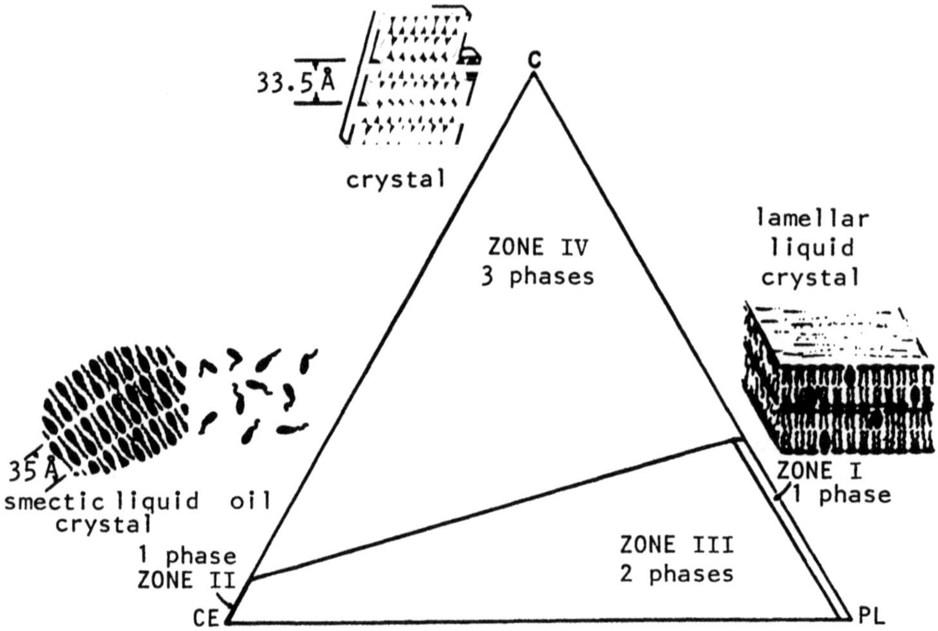

Fig. 9. Phase diagram of three-component system, cholesterol (C), phospholipid (PL), cholesteryl ester (CE), in excess water at 37°C and 1 atmosphere pressure. Zone 1 has a single phase of phospholipid lamellar liquid crystal with up to 33% cholesterol and 2% cholesteryl ester. Zone II also has a single phase composed of liquid or liquid crystalline cholesteryl ester. Mixtures in zone III have both the cholesteryl ester and the phospholipid phase present. In zone IV there is a third phase as well, cholesterol monohydrate crystals. Schematic molecular representation of the phases shown near each apex of the triangle. Molecules of cholesteryl ester form isotropic oil droplets when random, and birefringent droplets when ordered in layers 35 A thick. (From Katz _et al._[61])

the continuous one. This result arises because the newly formed interfaces of the small droplets of the disperse phase produced during emulsification are not immediately in equilibrium with the surrounding continuous phase. Due to a relative scarcity of surfactant in the thin film of continuous liquid between two juxtaposed droplets, less adsorption occurs in this region and the interfacial free energies are higher than in other regions of the droplets which are exposed to the open solution and where diffusion and adsorption of surfactant can occur freely. The resultant inter-

Table 4

The Effect of Mixtures of Lecithin and Cholesterol on Triglyceride Oil Emulsions[a]

Ratio Lecithin/Cholesterol (w/w)	Emulsion Type
19	O/W
10	O/W
8	W/O/W (duplex)
6	W/O
4	W/O
2	W/O

a. These data for 50% v/v olive oil and water emulsions are taken from Corran[61].

facial tension gradients cause a flow of surfactant and underlying continuous phase into the thin film region so that coalescence is inhibited and the droplets are stabilised. Adsorption from the interior of the dispersed droplet would not result in a surface tension gradient; thus no stabilisation of freshly formed droplets occurs and phase inversion of the emulsion can occur. Such destabilisation of the O/W systems can occur when $\lesssim$ 8/1 w/w lecithin/cholesterol is used as the emulsifier (Table 4). Since the phosphatidylcholine/cholesterol ratios in lipoproteins are less than this value (Table 1), the cholesteryl ester and triglyceride in serum lipoproteins could not be satisfactorily emulsified into water because of the antagonistic effects of phosphatidylcholine and cholesterol. Of course, since the phase volume of fat in serum is only ~ 1% complete inversion is not possible but local destabilisation could occur. The above reasoning therefore suggests that the apoprotein present at the surface of lipoprotein particles is necessary to stabilise the O/W systems.

V. SUMMARY AND CONCLUSIONS

The physical properties of cholesterol, cholesteryl esters, triglycerides, phospholipids and their interactions with water are considered. Hydrocarbon chains adopt the all trans configuration in crystals and lipids can exist in different polymorphic forms because of variations in the angle of chain packing. Trans-gauche isomerisations occur in the chains on heating and a cooperative melting transition takes place at a characteristic temperature; the cooperativity in HDL and LDL particles is low because of their small size. Cholesteryl esters and phospholipids do not only melt to the isotropic liquid but can form liquid crystalline phases in which the molecular order is greater than in the isotropic liquid. Lipids are surface active and associate readily in water because of the hydrophobic effect; the hydrophobic free energy is proportional to the length of the n-alkane chain. The phase behaviour of hydrated lipid systems is a function of the shape and charge of the lipid molecules. While liquid triglycerides and cholesteryl esters mix to form water-immiscible oils, phospholipids tend to form separate structures because the polar groups interact strongly with water. The apoproteins in lipoproteins are probably necessary to stabilise the particles in the aqueous phase because phosphatidylcholine and cholesterol have antagonistic emulsifying actions so that, at the ratios which exist in lipoproteins, they cannot stabilise an oil-in-water emulsion satisfactorily.

REFERENCES

1. R.I. Levy, D.W. Bilheimer and S. Eisenberg, in "Plasma Lipoproteins", ed. R.M.S. Smellie, Biochem. Soc. Symp. No. 33, Academic Press, London, 1971, pp. 3-17.
2. A.M. Scanu, in "Atherogenesis: Initiating Factors", Elsevier, Amsterdam, 1973, pp. 223-249.
3. M.C. Phillips, Prog. Surface Membrane Sci. 5 (1972) 139-221.
4. A.G. Lee, Prog. Biophys. Mol. Biol. 29 (1975) 3-56.
5. G. Lagaly, Angew. Chem. 15 (1976) 575-586.
6. B.D. Ladbrooke and D. Chapman, Chem. Phys. Lipids 3 (1969) 304-356.
7. J.M. Sturtevant, in "Quantum Statistical Mechanics in the Natural Sciences", ed. B. Kursonoglu, S. Mintz and S. Widmayer, Plenum Press, New York, 1974, pp. 63-84.
8. H.J. Hinz and J.M. Sturtevant, J. Biol. Chem. 247 (1972) 6071-6075.
9. D. Marsh, A. Watts and P.F. Knowles, Biochim. Biophys. Acta 465 (1977) 500-514.
10. D.G. Dervichian, Prog. Biophys. Mol. Biol. 14 (1964) 263-342.
11. D. Chapman, "The Structure of Lipids", Methuen, London, 1965.
12. V. Luzzati, in "Biological Membranes", ed. D. Chapman, Academic Press, New York, 1968, pp. 71-123.
13. R.M. Williams and D. Chapman, Prog. Chem. Fats & Other Lipids 11 (1970) 1-79.
14. A. Tardieu, V. Luzzati and F.C. Reman, J. Mol. Biol. 75 (1973) 711-733.
15. Hitchcock, P.B., Mason, R., Thomas, K.M. and Shipley, G.G. Proc. Nat. Acad. Sci. USA 71 (1974) 3036-3040.
16. Hitchcock, P.B., Mason, R. and Shipley, G.G., J. Mol. Biol. 94 (1975) 297-299.
17. Y.K. Levine, Prog. Biophys. Mol. Biol. 24 (1972) 3-74.
18. G.G. Shipley, in "Biological Membranes Vol. 2", ed. D. Chapman and D.F.H. Wallach, Academic Press, New York, 1973, pp. 1-89.
19. B.M. Craven, Nature 260 (1976) 727-729.
20. B.M. Craven and G.T. DeTitta, J. Chem. Soc. Perkin II (1976) 814-822.

21. K. Van Putte, W. Skoda and M. Petroni, Chem. Phys. Lipids 2 (1968) 361-371.
22. G.W. Gray, "Molecular Structure and the Properties of Liquid Crystals", Academic Press, New York, 1962.
23. D.M. Small, C. Loomis, M. Janiak and G.G. Shipley, in "Liquid Crystals and Ordered Fluids", ed. J.F. Johnson and R.S. Porter, Plenum Press, New York, 1974, Vol. 2, pp. 11-22
24. D. Chapman, Science J. (1965) 32-38.
25. D. Chapman, in "Biological Membranes", ed. D. Chapman, Academic Press, New York, 1968, pp. 125-202.
26. H. Hauser, in "Water. A Comprehensive Treatise", ed. F. Franks, Plenum Press, New York, 1975 Vol. 4, pp. 209-303.
27. M.C. Phillips, R.M. Williams and D. Chapman, Chem. Phys. Lipids 3 (1969) 234-244.
28. J.E. Rothman, J. Theor. Biol. 38 (1973) 1-16.
29. A.G. Lee, N.J.M. Birdsall, Y.K. Levine and J.C. Metcalfe, Biochim. Biophys. Acta 255 (1972) 43-56.
30. A. Bondi, Chem. Rev. 67 (1967) 565-580.
31. E.M. Barrall and J.F. Johnson, in "Liquid Crystals and Plastic Crystals", Vol. 2, ed. G.W. Gray and P.A. Winsor, Horwood, Chichester, 1974, pp. 254-306.
32. D.M. Small, in "Surface Chemistry of Biological Systems", ed. M. Blank, Plenum Press, New York, 1970, pp. 55-83.
33. G.J. Davis, R.S. Porter, J.W. Steiner and D.M. Small, Mol. Cryst. Liquid Cryst. 10 (1970) 331-336.
34. D. Atkinson, R.J. Deckelbaum, D.M. Small, and G.G. Shipley, Proc. Nat. Acad. Sci. USA 74 (1977) 1042-1046.
35. C. Tanford, "The Hydrophobic Effect", Wiley, New York, 1973.
36. F. Franks, in "Water. A Comprehensive Treatise", ed. F. Franks Plenum Press, New York, Vol. 4, 1975, pp. 1-94.
37. R.B. Hermann, J. Phys. Chem. 76 (1972) 2754-2759.
38. J.A. Reynolds, D.B. Gilbert and C. Tanford, Proc. Nat. Acad. Sci. USA 71 (1974) 2925-2927.
39. J.T. Davies and E.K. Rideal, in "Interfacial Phenomena", 2nd ed., Academic Press, London, 1963, p. 371.
40. M.N. Jones, in "Biological Interfaces", Elsevier, Amsterdam, 1975, p. 69.
41. R. Smith and C. Tanford, J. Mol. Biol. 67 (1972) 75-83.
42. M.E. Haberland and J.A. Reynolds, Proc. Nat. Acad. Sci. USA 70 (1973) 2313-2316.
43. D.G. Dervichian, Mol. Cryst. 2 (1966) 55-62.
44. D.M. Small, Fed. Proc. 29 (1970) 1320-1326.
45. A.S.C. Lawrence, Mol. Cryst. Liq. Cryst. 7 (1969) 1-57.
46. V. Luzzati and A. Tardieu, Ann. Rev. Phys. Chem. 25 (1974) 79-94.
47. J.L. Ranck, L. Mateu, D.M. Sadler, A. Tardieu, T. Gulik-Krzywicki and V. Luzzati, J. Mol. Biol. 85 (1974) 249-277.
48. D.M. Small, J. Lipid Res. 8 (1967) 551-557.
49. F. Reiss-Husson, J. Mol. Biol. 25 (1967) 363-381.

50. G.G. Shipley, L.S. Avecilla and D.M. Small, J. Lipid. Res. 15 (1974) 124-131.
51. H. Hauser and M.C. Phillips, Prog. Surface Membrane Sci., in press.
52. T. Gulik-Krzywicki, A. Tardieu and V. Luzzati, Mol. Cryst. Liq. Cryst. 8 (1969) 285-291.
53. C.R. Loomis, M.J. Janiak, D.M. Small and G.G. Shipley, J. Mol. Biol. 86 (1974) 309-324.
54. B. Lundberg, Acta Chem. Scand. B30 (1976) 150-156.
S. Ekman and B. Lundberg, Acta Chem. Scand. B30 (1976) 825-830.
55. M.J. Janiak, C.R. Loomis, G.G. Shipley and D.M. Small, J. Mol. Biol. 86 (1974) 325-339.
56. R.J. Deckelbaum, G.G. Shipley and D.M. Small, J. Biol. Chem. 252 (1977) 744-754.
57. W.D. Bancroft, J. Phys. Chem. 17 (1913) 501-519.
58. A.E. Bailey, in "Melting and Solidification of Fats", Interscience, New York, 1950, p. 172.
59. P. Mukerjee and K.J. Mysels, "Critical Micelle Concentrations of Aqueous Surfactant Systems", National Bureau of Standards Publications, Washington, 1971.
60. M. Hayashi, M. Okazaki and I. Hara, Proc. 6th Internat. Congress Surface-Activity, Carl Hanser, Munich, Vol. 2, 1973, 361-370.
61. J.W. Corran, in "Emulsion Technology", 2nd ed., Chemical Publishing Co., Brooklyn, 1946, pp. 176-192.
62. S.S. Katz, G.G. Shipley and D.M. Small, J. Clin. Invest. 58 (1976) 200-211.

THE USE OF PARAMAGNETIC PROBES IN THE STRUCTURE DETERMINATION OF HIGH DENSITY SERUM LIPOPROTEINS

H. Hauser

Eidgenössische Technische Hochschule, Zürich

Laboratorium für Biochemie, Universitätstr. 16,
CH-8006 Zürich, Switzerland

Introduction

The subject of this paper is the structure of high density lipoproteins (HDL) with particular reference to the nature of the lipid - protein interaction. The question of whether or not there are any general principles underlying the lipid - protein interaction in serum lipoproteins is of prime importance and will be discussed here. The work to be described was carried out with the HDL_3 fraction isolated from porcine plasma and characterized by an operationally defined density of 1.12 - 1.21 g/ml. Results obtained with the native system will be compared with those of a model system reconstituted from lecithin and apoproteins isolated from porcine HDL_3 lipoprotein. Several papers have discussed the striking compositional and structural similarities of human and porcine lipoproteins (1-4), and it is believed that the principles governing the lipid - protein interaction are generally applicable. To support this some experiments were repeated with human HDL of an operationally defined density d = 1.063 - 1.21 g/ml. The physical techniques used to derive the structural information to be discussed here are mainly NMR and ESR spectroscopy. Because of the small size of the high density lipoprotein particles (see below) the tumbling rate is sufficiently fast as to average out any residual dipolar broadening of NMR lines. Hence good high resolution spectra are obtained from high density lipoproteins.

1. NATIVE HIGH DENSITY LIPOPROTEINS

The structure (morphology) of high-density lipoprotein (HDL_3). Small-angle X-ray scattering.

The form of the small-angle X-ray scattering curve (Fig. 1, from the work of Atkinson et al. (5)) and the two well-resolved secondary maxima are typical for a highly symmetrical structure and/or a high degree of particle homogeneity. The small-angle region ($h<0.38\ nm^{-1}$)

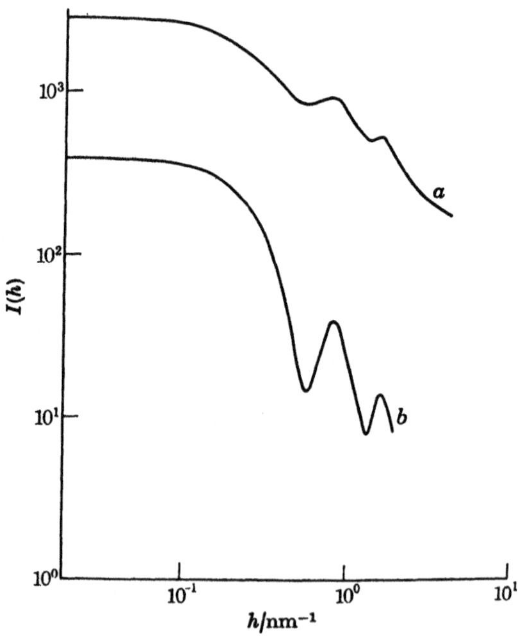

Fig.1. Small-angle X-ray scattering curve from HDL_3 in 0.15 M NaCl, 1 mM EDTA, 0.02% NaN_3 at infinite dilution.
(a) containing the influence of the line collimation system
(b) "desmeared" curve, after correction for these effects. I (h) in arbitrary units (from ref. 5).

was analyzed according to the treatment of Guinier yielding a radius of gyration R_g = 5.5 nm. The value of R_g and the position and relative intensities of the subsidiary scattering maxima are not consistent with a symmetrical particle of uniform electron density. From a comparison of the experimental scattering curve with theoretical curves computed for models consisting of a lipid core of low electron density and an electron dense outer shell the following conclusions were derived (5): HDL_3 is a spherical particle with a radius of gyration of 5.5 nm consisting of a central, low electron density (lipid) core of radius 4.2 nm and an outer electron dense, polar layer of a thickness of about 1.2 nm. The thickness of the electron dense outer shell of about 1 nm is consistent with the length of the polar group of phosphatidylcholine or a typical diameter of a protein α-helix suggesting that the outer layer may consist of phospholipid polar groups and protein α-helices. These features are summarized in Fig. 2. The radius of gyration of 5.5 nm is in good agreement with the Stokes radius of 5.6 nm determined by gel filtration on Sepharose 4B and with the apparent radius of 5.3 nm determined from electron micrographs of negatively stained HDL_3 samples.

The X-ray scattering at wider angles showed a single broad peak at an equivalent Bragg spacing of 0.45 nm characteristic of highly mobile, liquid crystalline hydrocarbon chains (5).

NMR results (6,11)

Fig. 3a and b are 270 MHz ^1H-NMR spectra of a sonicated egg lecithin dispersion and of HDL_3, respectively (6). From a comparison of the two spectra it is clear that (1) the lipoprotein spectrum is essentially a lecithin spectrum and (2) that, apart from a few resonances at 7.2 ppm (from 3-(trimethylsilyl)propane sulphonate) arising from aromatic protons and probably at 2.2 ppm no protein signals are visible. 360 MHz ^1H-NMR spectra of human HDL (d = 1.063 - 1.21 g/ml) closely resemble the spectrum shown in Fig. 3b (Hauser, H. and Kostner, G., to be published). Similar to the ^1H-NMR spectra most of the prominent resonances of the ^{13}C-spectra of both porcine HDL_3 and human HDL can be assigned to the

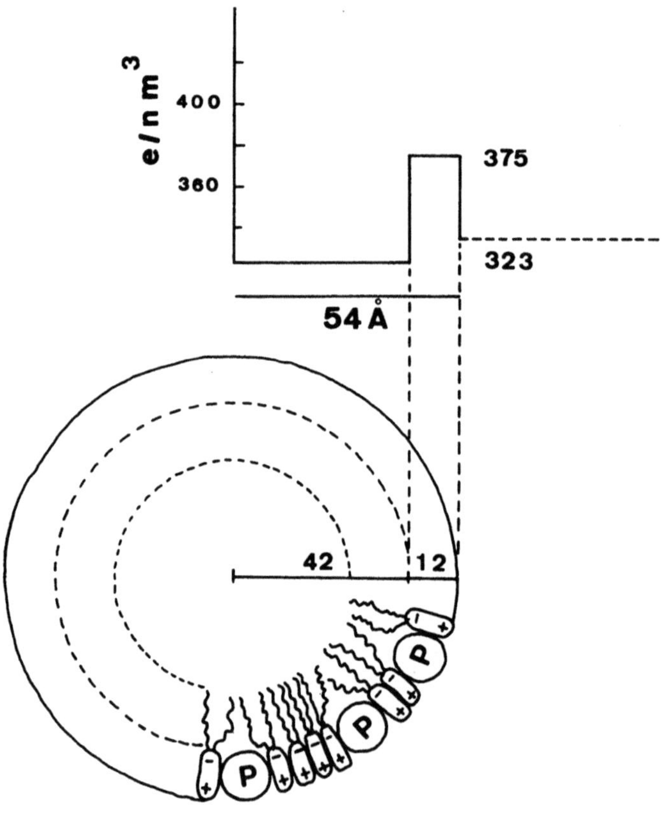

Fig. 2. Summary of the small-angle X-ray scattering data of HDL_3. The spherical particle of a radius of 5.4 nm is surrounded by an electron dense layer (375 e/nm^3) of 1.2 nm consisting of the phospholipid polar groups and the protein (P) a large proportion of which is present as amphipathic α-helical segments (shown in cross section (P)). The axes of the α-helical segments intercalated between the lipid polar groups run approx. perpendicular to the phospholipid hydrocarbon chains. The core of the particle (4.2 nm) consists of lipids, mainly cholesteryl esters, having a low electron density of 323 e/nm^3.

Any paramagnetic ion has a local magnetic field

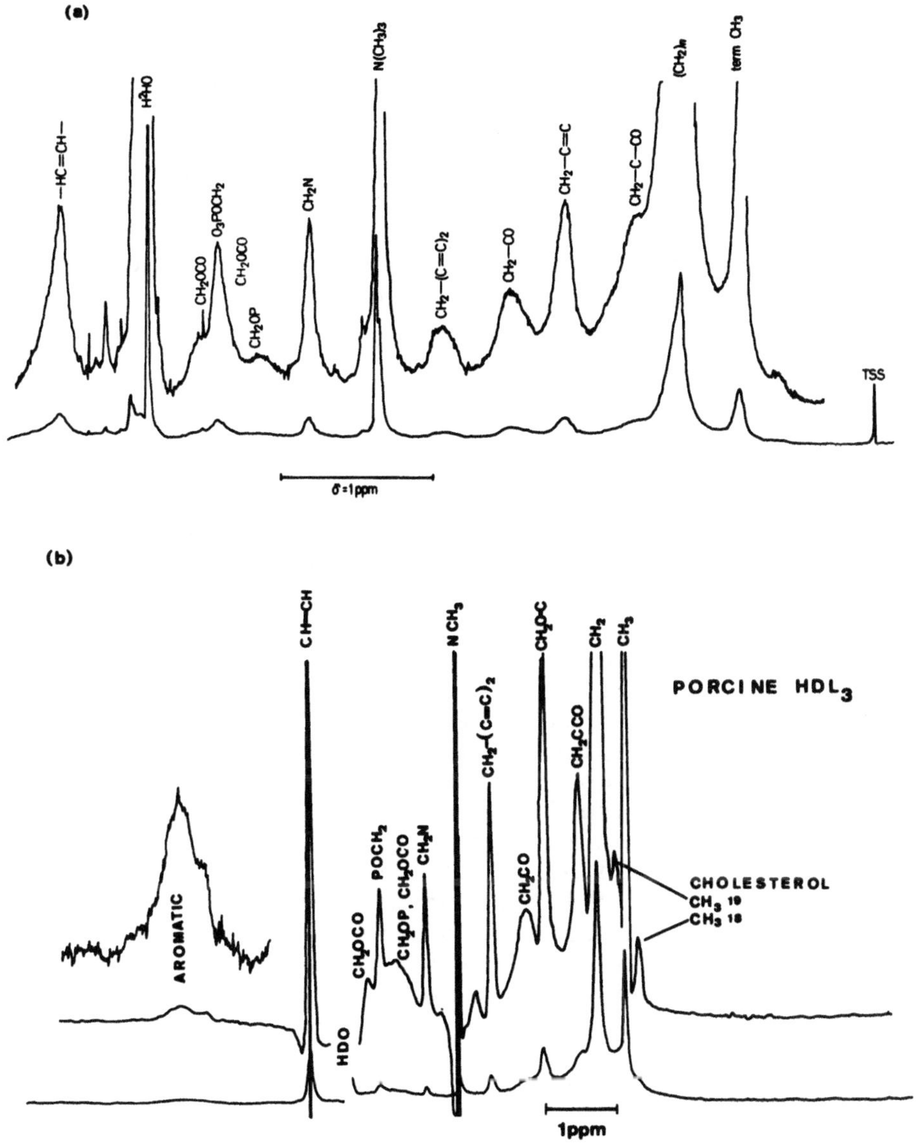

Fig. 3. (a) 270 MHz proton n.m.r. spectrum of a sonicated egg lecithin dispersion (20 mg/ml ≃ 0.027 M) in ^2H_2O (nominal pH 5.5). (b) 270 MHz proton n.m.r. spectrum of native porcine high density lipoprotein HDL_3 (d = 1.12-1.21 g/cm^3) in ^2H_2O (concentration 100 mg/ml) containing 0.15 M NaCl, 1 mM EDTA and 0.02% NaN_3 adjusted to a nominal pH 8.6.

lipid moiety (cf. 14).

The lack of protein resonances indicates that the protein is rather rigid having a high degree of secondary and/or tertiary structure. The NMR results are consistent with CD data showing that ~70% of the protein in HDL_3 is α-helical (9).

Chemical shift measurements (relative to trimethylsilyl-propane sulphonate (TSS) as an internal standard) on both porcine HDL_3 and human HDL showed that the resonances from the glycerophosphorylcholine group and the first few CH_2 groups of the hydrocarbon chains are shifted relative to the chemical shifts of these resonances in pure egg lecithin. Intensity measurements show that all the $N(CH_3)_3$ groups contribute to the ^1H spectrum.

Paramagnetic ions as structural probes

Any paramagnetic ion has a local magnetic field around its centre which can induce considerable perturbation in both chemical shifts and relaxation times of nearby nuclei. The paramagnetic ions used in structural (conformational) studies may often be classified as "shift probe" or "relaxation probe" depending on the type of perturbation induced. The former type of probe induces mainly changes in chemical shifts although the shifted resonances usually show some line broadening, the latter affects primarily the relaxation behaviour and causes line broadening. Here we use the paramagnetic anion $Fe(CN)_6^{3-}$ as a shift probe and the paramagnetic lanthanide Gd^{3+} as a relaxation probe to derive structural information on both native high density lipoproteins and on a reconstituted model system. Gd^{3+} is an effective relaxation probe because the cation has a symmetric electron distribution (f^7) which ensures that effects on chemical shift are negligible.

The addition of paramagnetic $K_3Fe(CN)_6$ to porcine HDL_3 and human HDL produced changes Δδ in the ^1H-chemical shifts of the polar group resonances. 100% of the $N(CH_3)_3$ signal was shifted upfield indicating that all the $N(CH_3)_3$ groups (including those of sphingomyelin) are present at the surface of the HDL-particle readily accessible to the $Fe(CN)_6^{3-}$ anion. Fig. 4 shows the shift changes $\Delta\delta$ of the $N(CH_3)_3$ signal of human HDL as a function of the molar ratio $K_3Fe(CN)_6$ / phospholipid (Hauser, H. and Kostner, G., to be published). The shift

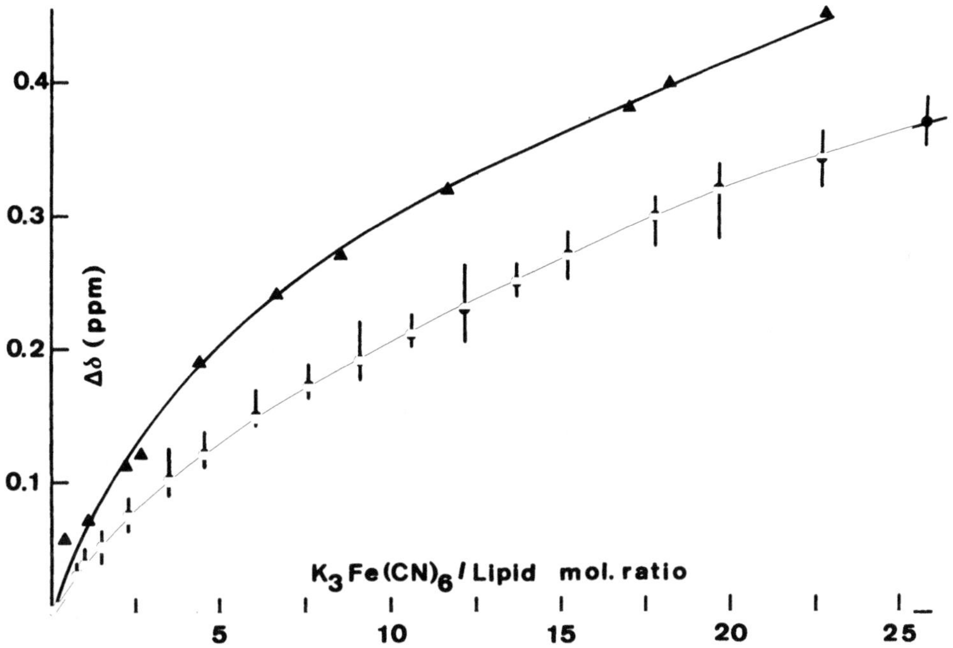

Fig. 4. Changes in chemical shift ($\Delta\delta$) of the ^1H $N(CH_3)_3$ signal as a function of the molar ratio $K_3Fe(CN)_6$ / lipid.
(▲—▲) human HDL (d= 1.063-1.21 g/ml)
(●—●) sonicated egg lecithin dispersion or sonicated phospholipid dispersion (composition: 70% egg lecithin, 20% sphingomyelin and 10% egg lysolecithin). Data were obtained at 360 MHz. The effective lipid concentration, i.e. the concentration of lipid effectively in contact with the $Fe(CN)_6^{-3}$ was 0.0176 M. In case of HDL the effective lipid concentration is identical to the total lipid concentration, in the case of the lipid dispersion the total concentration is 0.0267 M.

changes of the $N(CH_3)_3$ signal of sonicated lecithin and mixed phospholipid dispersions are shown for comparison. The changes in chemical shift $\Delta\delta$ of the other polar group signals were in the order $N(CH_3) > CH_2N > POCH_2 >$ (choline) $> CH_2OP$ (glycerol). No changes in chemical shift of the ^1H-signals from the lipid hydrocarbon chains and of the two ^1H-signals from cholesteryl ester and/or free cholesterol were observed up to $K_3Fe(CN)_6$ concentrations of 0.7M. These results are consistent with the schematic drawing in Fig. 2 showing that all phospholipids are arranged as a monomolecular surface layer surrounding the hydrophobic core. The location of the cholesteryl esters and free cholesterol is not conclusive. To solve this problem we intend to use the spectral perturbation in the ^{13}C-NMR spectrum induced by paramagnetic lanthanides.

The difference in the shift changes $\Delta\delta$ between lipoprotein and lipid dispersions (Fig. 4) is attributed to the presence of protein in the surface of the lipoprotein particle affecting the binding and/or accessibility of the $Fe(CN)_6^{3-}$ anion (diameter approx. 0.9 nm). The surface location of the protein is also apparent from the line width of the shifted $N(CH_3)_3$ resonance. In the presence of $K_3Fe(CN)_6$ the $N(CH_3)_3$ signal of HDL was broadened by a factor of 3-4 (at 360 MHz) compared to the $N(CH_3)_3$ signal in lecithin. The field dependence of the width (cf. 11) indicates that the broadening is due to a spread in chemical shifts, i.e. the $N(CH_3)_3$ signals are not equally shifted by $Fe(CN)_6^{3-}$. This reflects differences in the chemical environment and / or accessibility to $Fe(CN)_6^{3-}$ of the $N(CH_3)_3$ groups arising from the orientation of the protein in the surface of the lipoprotein particle. Exchange of lipid molecules between these different environments is slow on the NMR time scale (less than 10/s). Such slow exchange is consistent with the protein being intercalated between patches of lipid monolayer (Fig. 2) giving rise to a mosaic surface structure of the lipoprotein particle as discussed before (6). That the accessibility of the phospholipid polar group extends to cations is evident from ^{31}P-NMR experiments. The ^{31}P-NMR spectra of human HDL consist of three chemically shifted signals which, on the basis of chemical shift and intensity measurements can be assigned to lecithin, lysolecithin and sphingomyelin, respectively (cf. 15). Within the error of the measurement the spectral intensities were in agreement with the phospholipid composition. Adding increasing quantities of $Gd(NO_3)_3$ caused broadening of all three resonances whereby the total signal intensities were affected. This is consistent with the $Fe(CN)_6^{3-}$ experiments indicating that all the phospholipids are arranged in a monolayer on the outer surface of the lipoprotein particle (Fig. 2).

Comparing line width and T_1 measurements of porcine HDL_3 (11) with corresponding measurements on lecithin bilayers it can be concluded that the presence of the protein does not change significantly the molecular motion of the lipid except that the anisotropy of motion is increased (Hauser, H. and Kostner, G., to be published).

2. MODEL SYSTEM

The results obtained with the native HDL_3 are compared with a model system consisting of dimyristoyl-L-3-lecithin and apoprotein extracted from porcine HDL_3. The lipoprotein (lipid/protein molar (weight) ratio = 100 (2.5)) was prepared by mixing the unsonicated lipid dispersion with the appropriate amount of protein and subjecting the mixture to short bursts of ultrasonication. The above molar ratio represents the minimum amount of lipid required to completely complex one apoprotein molecule as determined by density gradient centrifugation (10).

Small-angle X-ray scattering (7) shows that the reconstituted system differs from the spherical geometry of the native HDL_3 system. The complex is an oblate ellipsoid with 11.0 nm as the major axis and 5.5 nm as the minor axis. The oblate shape is dictated by the fundamental bilayer organisation of the lecithin being retained (see discussion below). Consistent with that is the observation of a sharp diffraction at an equivalent Bragg spacing of 0.42 nm at temperatures below 30°C and the transition of this diffraction to a broad scattering peak at an equivalent Bragg spacing of 0.45 nm at temperatures >30°C. Such a transition is characteristic of the crystal-to-liquid crystal transition in phospholipid bilayers. Using synthetic lecithins differing in hydrocarbon chain length it can be shown that the dimension of the minor axis correlates with the length of the hy-

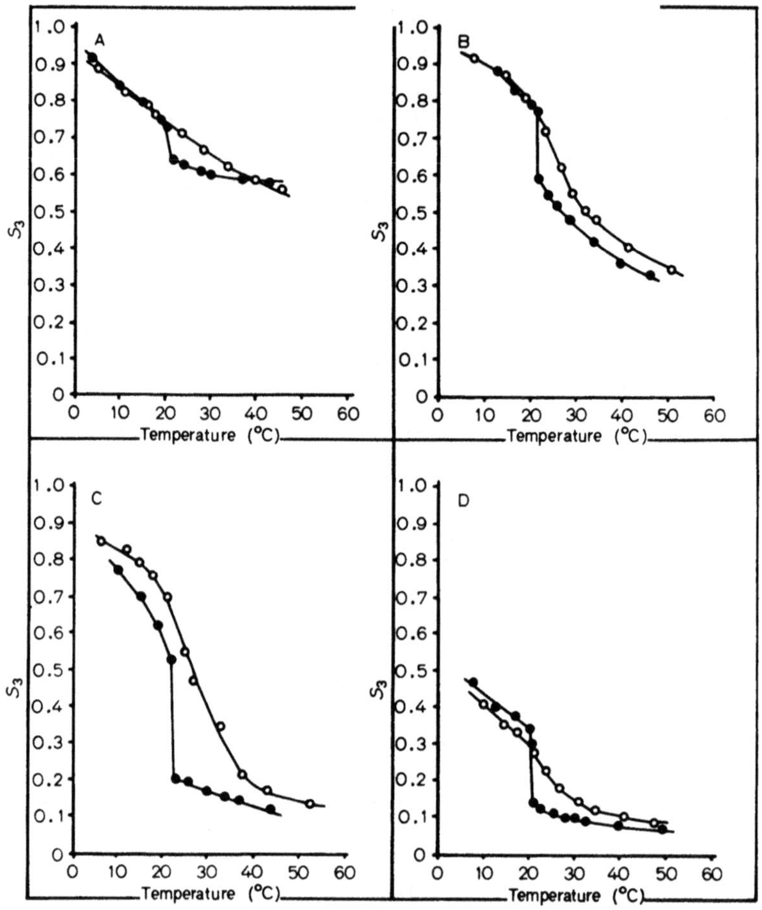

Fig. 5. Order parameters (S_3) as a function of temperature. Dimyristoyl lecithin dispersions (-●-) and the model lipoprotein (-○-) reconstituted from dimyristoyl lecithin and apoprotein extracted from porcine HDL_3 were spin-labelled with the following fatty acid spin labels I (m,n);

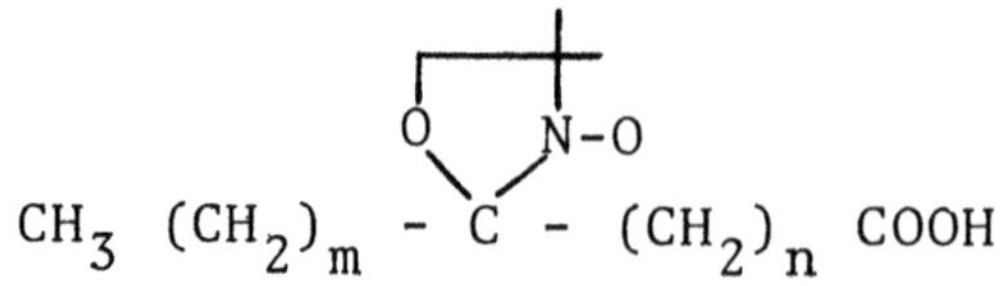

(a) I (12,3) pH 7.5 (5 mM Tris, 1 mM EDTA, 50 mM NaCl, 0.02% NaN_3)
(b) as (a) but pH 3.5 adjusted by the addition of HCl.
(c) I (5.10) pH 7.5
(d) I (1,14) pH 7.5
From ref. 8.

drocarbon chains which are oriented perpendicular to the interface. The difference in the morphology between native and reconstituted system is probably due to the absence of cholesteryl esters in the model. Further evidence for the phospholipid being present as a bilayer is furnished independently by ESR spin label experiments and differential scanning calorimetry. Fig. 5 shows that all spin probes incorporated in dimyristoyl lecithin bilayers report sharp chain-melting transitions between 20 and 23°C. With pure lecithin bilayers at temperatures above the chain-melting transition the order parameters as defined in ref. (12) showed a steady decrease with increasing distance of the nitroxide radical from the lipid - water interface. This so-called flexibility gradient typical for phospholipid bilayers is shown in Fig. 6. In the reconstituted system the labels with the nitroxide group close to the centre of the bilayer showed a broadened phase transition while with the fatty acid having the nitroxide in position 5 no chain-melting transition was observed indicating that the prime site of interaction of the apoprotein is the glycerylphosphorylcholine group of lecithin. Furthermore, all three fatty acid spin labels used showed higher order parameters in the complex than in the pure bilayer although the shape of the flexibility gradient in the reconstituted lipoprotein system was identical to that measured with pure lecithin bilayers (Fig. 6). This indicates that the lipid is not immobilized as a result of the interaction with the apoprotein and therefore strong interactions between the lipid and the protein can be ruled out. In the complex the lipid motion is more anisotropic as indicated by the higher order parameter. Consistent with the ESR spin label results are differential scanning calorimetry data showing a main endothermic peak at 23°C typical for dimyristoyl lecithin bilayers.

The NMR results obtained with the reconstituted system (cf. 8) are consistent with the data reported for the native lipoproteins indicating that the mode of the lipid-protein interaction is similar in the two systems. This is remarkable considering the difference in the morphology of the two systems.

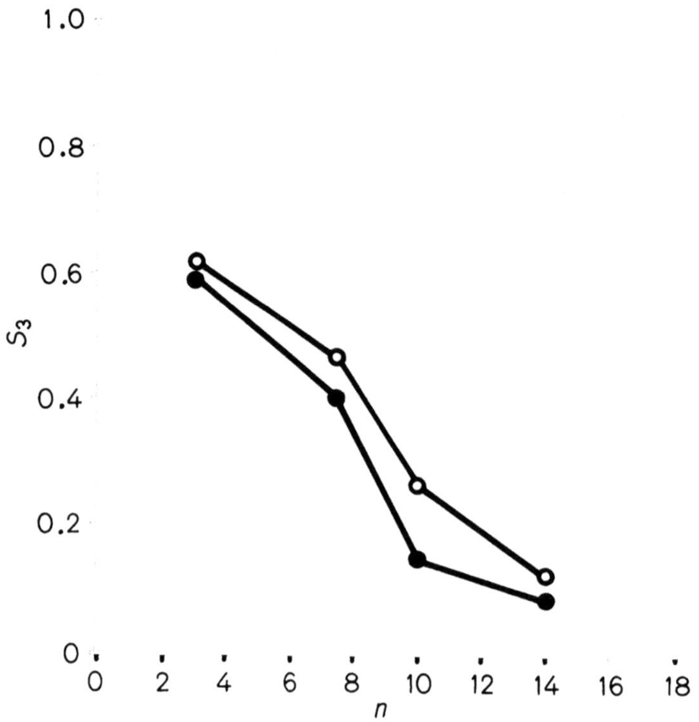

Fig. 6. Order parameter S_3 at 35°C as a function of the position of the nitroxide group along the hydrocarbon chain; dimyristoyl lecithin bilayers (-●-); lipoprotein complex reconstituted from dimyristoyl lecithin and porcine HDL_3-apoprotein (-o-). From ref. 8.

3. SUMMARY

Comparing the native HDL with the reconstituted system it is concluded that despite morphological differences the principles governing the lipid-apoprotein interaction are similar in both systems. The lipid-protein interaction is characterized as follows:

(1) The prime site of the relatively weak lipid-protein interaction is in the polar region of the phospholipid, i.e. the glycerylphosphorylcholine group.

(2) The surface of the HDL-particle has a mosaic structure consisting of patches of phospholipid monolayer separated by protein. All phospholipids are oriented at the external surface as shown in Fig. 2 with their polar

groups readily accessible to ions in the aqueous phase.

(3) The lipid molecules are not immobilized by the interaction with the protein and therefore strong binding does not occur. The molecular motion of the lipid is similar to that observed in bilayers except that the interaction with the protein increases the anisotropy of motion.

(4) The NMR data are consistent with the protein being fairly rigid and most of the peptide segments being oriented at the lipid-water interface. Since the α-helix content is known to be high the axes of the α-helices must be running along the interface(perpendicular to the hydrocarbon chain axis).The high affinity of the protein for the lipid-water interface discussed before (13) and the specific interfacial arrangement of the peptide segments are probably responsible for the emulsifying properties of apoproteins.

References

(1) Davis, M.A.F., Henry, R. and Leslie, R.B. (1974) Comp. Biochem. Physiol., Vol. 47 B, pp 831-849.

(2) Jackson, R.L., Baker, H.N., Taunton, O.D., Smith, L.C., Garner, C.W. and Gotto, A.M. (1973) J. Biol. Chem. 248, 2639-2644.

(3) Mills, G.L. and Taylaur, C.E. (1971) Comp. Biochem. Physiol. 40 B, 289-301.

(4) Fidge, N. (1973) Biochim. Biophys. Acta 295, 258-273.

(5) Atkinson, D., Davis, M.A.F. and Leslie, R.B. (1974) Proc. R. Soc. Lond. B 186, 165-180.

(6) Hauser, H. (1975) Febs Letters 60, 71-75.

(7) Atkinson, D., Smith, H.M., Dickson, J. and Austin, J.P. (1976) Eur. J. Biochem. 64, 541-547.

(8) Andrews, A.L., Atkinson, D., Barratt, M.D., Finer, E.G., Hauser, H., Henry, R., Leslie, R.B., Owens, N.L., Phillips, M.C. and Robertson, R.N. (1976) Eur. J. Biochem. 64, 549-563.

(9) Morrisett, J.D., Jackson, R.L. and Gotto, A.M., Jr. (1975) Ann. Rev. Biochem. 44, 183-207.

(10) Hauser, H.,Henry, R., Leslie, R.B. and Stubbs, J.M. (1974) Eur. J. Biochem. 48, 583-594.

(11) Finer, E.G., Henry, R., Leslie, R.B. and Robertson, R.N. (1975) Biochim. Biophys. Acta 380, 320-337.

(12) Hubbel, W.L. and McConnell, H.M. (1971) J. Amer. Chem. Soc. 93, 314-326.

(13) Davis, M.A.F., Hauser, H., Leslie, R.B. and Phillips, M.C. (1973) Biochim. Biophys. Acta 317, 214-218.

(14) Hamilton, J.A., Talkowski, C., Childers, R.F., Williams, E., Allerhand, A. and Cordes, E.H. (1974) J. Biol. Chem. 249, 4872-4878.

(15) Assmann, G., Sokoloski, E.A. and Brewer, Jr., H.B. (1974) Proc.Nat.Acad.Sci. USA 71, 549-553.

REASSEMBLY OF APOPROTEINS AND LIPIDS

Maryvonne ROSSENEU

Simon Stevin Instituut voor Wetenschappelijk Onderzoek

Ziekenhuis Sint Jan, Ruddershove, B-8000 Brugge, Belgium

INTRODUCTION

An approach to the problem of lipid-protein interactions in lipoproteins, is to study the behaviour of "in vitro" reassembled apoprotein-lipid mixtures. Apoproteins obtained by delipidation of lipoproteins can be recombined with lipids either synthetic or isolated from the corresponding lipoprotein fraction. Analysis of the kinetics and thermodynamic features of the association process, as well as of the composition, properties and structure of the complex, should provide information about the nature and extent of the forces responsible for the formation and stabilization of the apoprotein-lipid bond within native lipoproteins. Various technical problems are associated with these experiments such as the preparation and characterization of lipid dispersions and of apoprotein solutions, the selection of an optimal reassembly procedure given the lipid and protein composition of the initial mixture. Discussion and evaluation of the various reassembly techniques will be the first part of this paper. The physical techniques applied to the monitoring of the extent of association and to the isolation and characterization of the lipid-apoprotein complexes will be subsequently described. Finally, some of the results obtained with isolated apoA-I, apoA-II, apoC-I, apoC-III proteins will be summarized and analysed.

I. PREPARATION OF THE LIPIDS FOR REASSEMBLY EXPERIMENTS

A. Dispersion of phospholipids

It seems now well established by many "in vitro" recombination studies, specially by LUX et al. (1), SCANU et al. (2,3), that binding of phospholipids is a prerequisite to the binding of neutral lipids to apoproteins. According to HAMILTON (4) nascent HDL appear as an apoA-II-phospholipid complex. This stresses and justifies the importance of the efforts to isolate and characterise phospholipid-apoprotein complexes.

The procedures to recombine apoproteins with phospholipids are dependent on the nature of the latter components. Among the phospholipids a first distinction should therefore be made between micellar phospholipids consisting of lysolecithins and short-chain lecithins, and apolar phospholipids consisting of long-chain lecithins and sphingomyelin (Fig. 1.).

1. Micellar phospholipids. Due to their amphipathic nature phospholipids have a tendency to self-associate into condensed molecular aggregates in aqueous media; in the case of lysolecithins and short-chain lecithins these aggregates are micelles, characterized by their aggregation number, critical micellar concentration, size and shape (5). These properties are determined by the length and degree of unsaturation of the acyl chains, ionic strength, etc. Lysolecithin and short-chain lecithins such as dihexanoyl, dioctanoyl, -L-alpha lecithins are water soluble and form optically clear solutions in water. They can also by used in the monomeric state since their critical micellar concentration is low (6,7). However, as short-chain lecithins are not present in lipoproteins and lysolecithin represents only a minor phospholipid constituent, these binding studies might not be significantly related to the biological conditions.

2. Lecithin and Sphingomyelin. Long-chain lecithins and sphingomyelin require special procedures for their preparation and isolation since they can exist in many physical forms (8,9). Anhydrous preparations of lecithins exhibit various physical states of aggregation as water is added. Such behaviour is called: lyotropic mesomorphism (10,11). When the content of water exceeds about 50% a smectic phase with a liquid-like hydrocarbon region appears. In the presence of an excess of water which is the case in all reassembly experiments carried out so far, lecithin and sphingomyelin associate to form molecular structures or liposomes, whose lamellar configuration is referred to as a "smectic mesophase". This structure can be visualized as stacked lamellae of phospholipid molecules in a bilayer configuration with the water taking up the space between the lamellae and interacting with the polar groups of phospholipids (12, 13). The tendency of phospholipids to assume this hydrated lamellar configuration is related to the temperature for the melting of hydrocarbon chains. Thus, natural unsaturated phospholipids, above this transition at ambient temperature, will form these structures spontaneously. Synthetic phospholipids with saturated acyl chains will also form similar structures if the temperature is raised close to the transition point of the dry compounds (14).

The original preparation of these multilamellar vesicles (15) was made by simple mechanical shaking of a dry phospholipid sample with a salt solution in a round-bottom flask. This method yields heterogeneous mixture of particles of varying size.

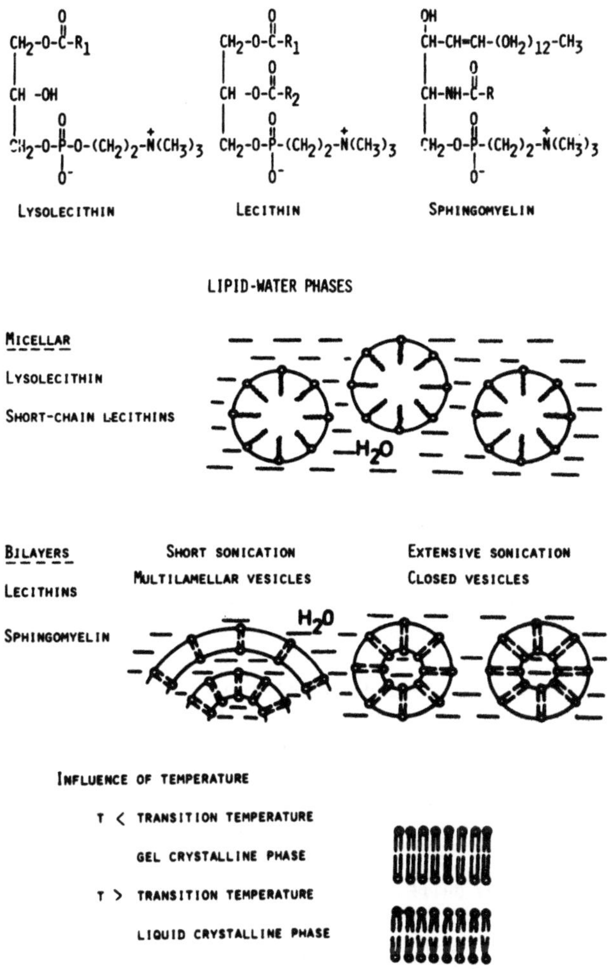

Fig. 1. Physical state of phospholipid dispersions

Exposure of the coarse phospholipid suspensions to ultrasonic radiation tends to break the large multilamellar vesicles into smaller ones and eventually produces a homogeneous population of unilamellar vesicles of 200-500Å diameter (16). Decrease of the size of the vesicles is accompanied by a progressive decrease of the turbidity of the phospholipid dispersion. This can be monitored by measuring the optical density at 450nm (Fig. 2.). After sonication at 37° and 50W of a dimyristoyl lecithin suspension for 15' the optical density remains constant at 0.1. The length of the sonication period depends on the sonication temperature and power. At this stage the suspension consists of more than 80% unilamellar lecithin vesicles.

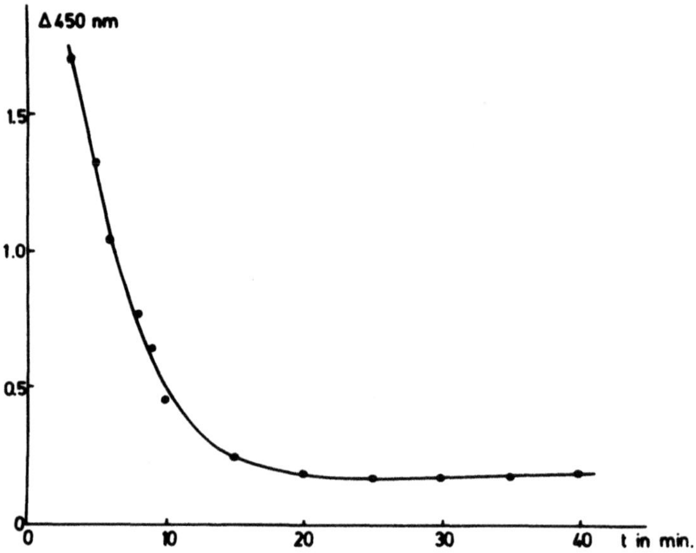

Fig. 2. Monitoring of the optical density decrease during sonication of dimyristoyl lecithin (37°, 50W)

Light scattering and sedimentation velocity studies of aqueous dispersions of egg lecithin vesicles indicate an aggregate weight of 1.5 to $2.1.10^6$ daltons, corresponding to approximately 1.9 to $2.7.10^3$ molecules per particle (16). Dimyristoyl lecithin and sphingomyelin vesicles can be fractionated by gel chromatography on a Sepharose 6B column into large vesicles eluting in the void volume and into closed single-shelled liposomes (Fig.3). According to the calibration of this column the molecular weight of the lecithin liposomes is $1.6.10^6$ and that of sphingomyelin is $3.0.10^6$ daltons.

The stability of these lipid aggregates is dependent on the method of preparation, time, temperature and chemical composition. According to SUURSKUSK et al. the transition temperature of small vesicles is lower than that of larger vesicles (17). As a function of time these vesicles tend to coalesce to form larger aggregates. To prevent this phenomenon they should preferably be kept at temperatures higher than their transition temperature.

We compared the enthalpy liberated on incubating small and large dimyristoyl lecithin vesicles with apoHDL at 25°, i.e. above the phospholipid transition temperature (t = 23°). The complex formation is more highly exothermal with small vesicles (7). The enthalpy is a function of both the temperature at which vesicles sonication occurs and apoprotein-vesicles reassembly takes place. At t > transition temperature, the enthalpy is not sensitive to the sonication temperature as phospholipids are in a fluid state. At a reassembly temperature < transition temperature, the energy liberated during the complex formation increases with the sonication temperature i.e. with the energetical state of the phospholipid dispersion (Fig. 4).

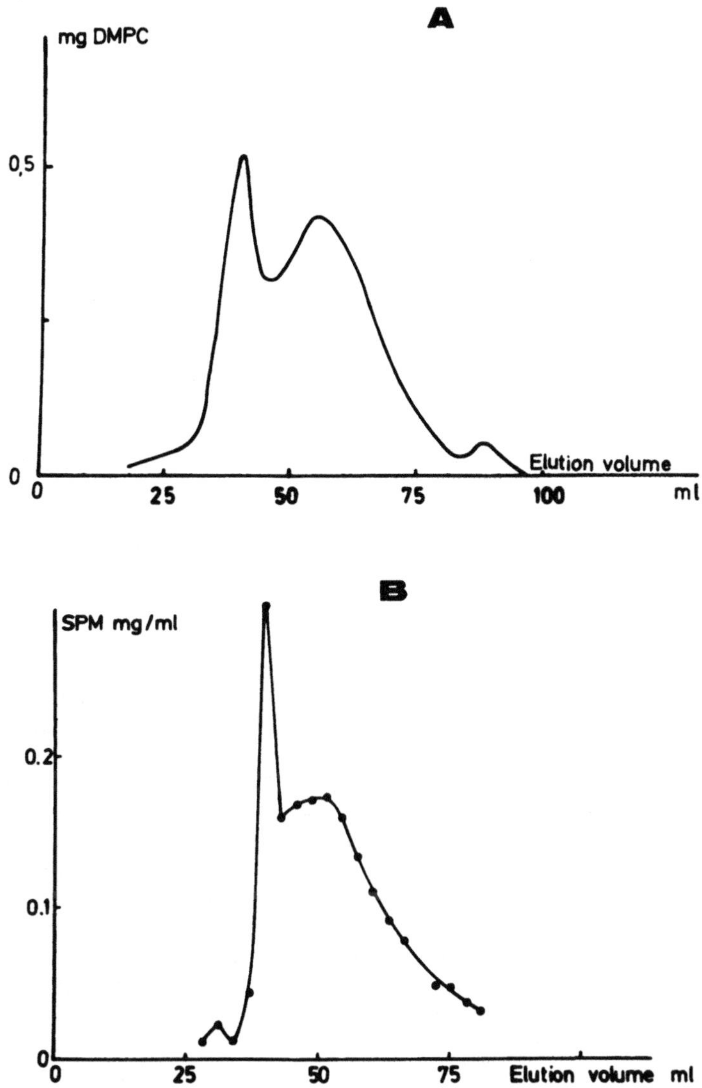

Fig. 3. Fractionation of lecithin (3A) and sphingomyelin (3B) liposomes on a Sepharose 6B column.

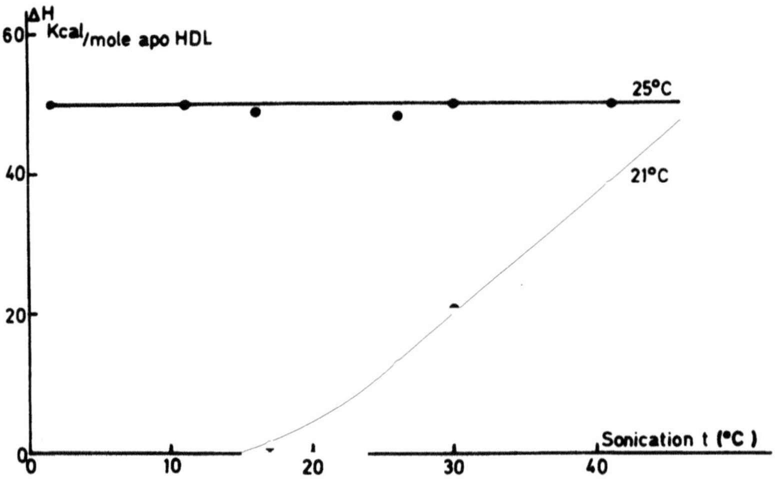

Fig. 4. Enthalpy change on binding dimyristoyl lecithin vesicles to human apoHDL. Reassembly temperatures were 25° and 21°, and the lecithin was sonicated at temperatures varying between 4 and 40°.

PHILLIPS et al. (18) have observed that the kinetics of complex formation between porcine apoHDL and multilamellar lecithin vesicles are slower than with single-shelled vesicles, as the protein interacts only with the outer bilayer. Eventually the same type of complex was obtained. Therefore closed unilamellar vesicles seem superior to the multilamellar ones for reassembly experiments with maximum yield and reproducibility of complex formation.

Another procedure for the preparation of unilamellar vesicles was proposed by BAZRI and KORN (19). It consists in flushing an ethanolic solution of the phospholipid into the aqueous phase by means of a microsyringe. The ethanol should be dialysed out of the vesicle dispersion afterwards and the vesicles concentrated on an Amicon X-100 membrane. Though this procedure yields small vesicles it is less suitable for calorimetric experiments as complete dialysis of the ethanol proved difficult and hard to reproduce. As a consequence the dilution heat of the vesicles was large and the presence of the organic solvent might interfere with protein binding. BRUNNER et al. (20) have recently proposed a method to prepare single-shelled lecithin-vesicles of uniform size by solubilizing unsonicated lecithin dispersions with sodium cholate and removing the detergent from the mixed lecithin-cholate micelles by gel filtration on Sephadex G-50. A homogeneous population of single-bilayer vesicles free of multilamellar structure is obtained by this method.

B. Dispersion of neutral lipids

At the present, there are no methods for preparing pure, homogeneous, triglyceride, cholesterol or cholesteryl ester particles for studying lipid-protein association. However when mixed with lecithin these neutral lipids can exist in well defined uniform structures. Cholesterol-lecithin form stable complexes up to a ratio of 2:1 (21). According to LUNDBERT and SAARINEN (22), stable lecithin-cholesterylester mixtures are obtained by cosonication at a minimal ratio of 0.5/1 (W:W). Fig. 5. shows that, after 30' cosonication at 40°, dimyristoyl lecithin and cholesterol oleate, above a 0.5/1 ratio exist as a stable dispersion with an optical density of 0.1.

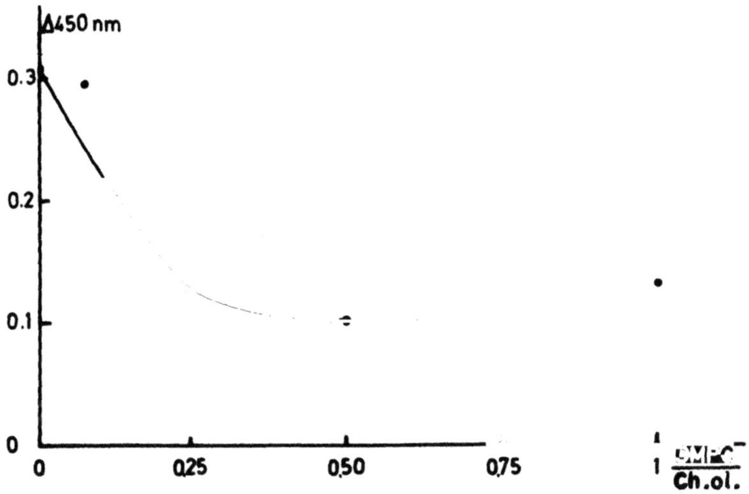

Fig. 5. Optical density of a dimyristoyl lecithin-cholesterol oleate dispersion after cosonication at 40°, 50W for 30'.

II. APOPROTEIN ISOLATION AND SELF ASSOCIATION

Procedures for the isolation and purification of apolipoproteins have been extensively described (23, 24, 25). As efficient isolation procedures imply the use of either detergents, urea or guadidium hydrochloride, this suggests that the polypeptides easily self-associate in aqueous buffer solutions. These processes are now well documented for the apoA-I and apoA-II proteins of human and non-human primates (26, 27, 28, 29, 30), suggesting that the lower polymers have a higher affinity for phospholipids (31).

Several agents are able to counteract the self-association process of the apoA-I protein, i.e. detergents such as sodium decyl- or dodecylsulfate (32), lysolecithin (31) or the presence of another apoprotein (33).

We observed that, after preincubation with lysolecithin, the enthalpy change evolved on binding the human apoA-I protein to dimyristoyl lecithin liposomes increases by -25Kcal/mole apoprotein. This difference corresponds to the enthalpy for dissociation and conformational changes of the apoA-I protein (31).

As a consequence, lipid-apoprotein reassembly studies which involve the presence of both lipid and protein aggregate should be performed under carefully controled conditions so as to yield reproducible results, which can be compared to those from other studies.

III. PROCEDURES FOR REASSEMBLING LIPIDS AND APOPROTEINS

Lipid-apoprotein complexes "in vitro" in bulk aqueous solutions have been obtained through two main types of procedure, the one requiring ultrasonic irradiation of the phospholipids only, while the other required an extra co-sonication step of the mixture. Both procedures will be described briefly.

A. Procedures with cosonication

As proposed by SCANU (34) a phospholipid dispersion in an aqueous buffer, is sonically irradiated in the presence of nitrogen for a fixed period of time, at a temperature above the melting point of the phospholipid acyl chains, in a container of fixed geometry. After addition of the purified protein, the mixture is further sonicated. Several authors followed this experimental procedure in earlier reassembly work (35).

We have stepwise reassembled the lipid and protein moiety of HDL, LDL and VLDL by cosonication in the following sequence : phospholipids, plus cholesterol, plus cholesterol esters, plus triglycerides, plus apoproteins. Each addition involved a 15' sonication step at 40°.

A comparison of the microviscosity of the reassembled mixture with that of native VLDL and HDL (Fig. 6A and B) shows that the values are very close, suggesting that the reassembly process yields an homogeneous mixture (36). For LDL the microviscosity of the complex remains lower, probably due to an incomplete solubilization of apoB (Fig. 6C).

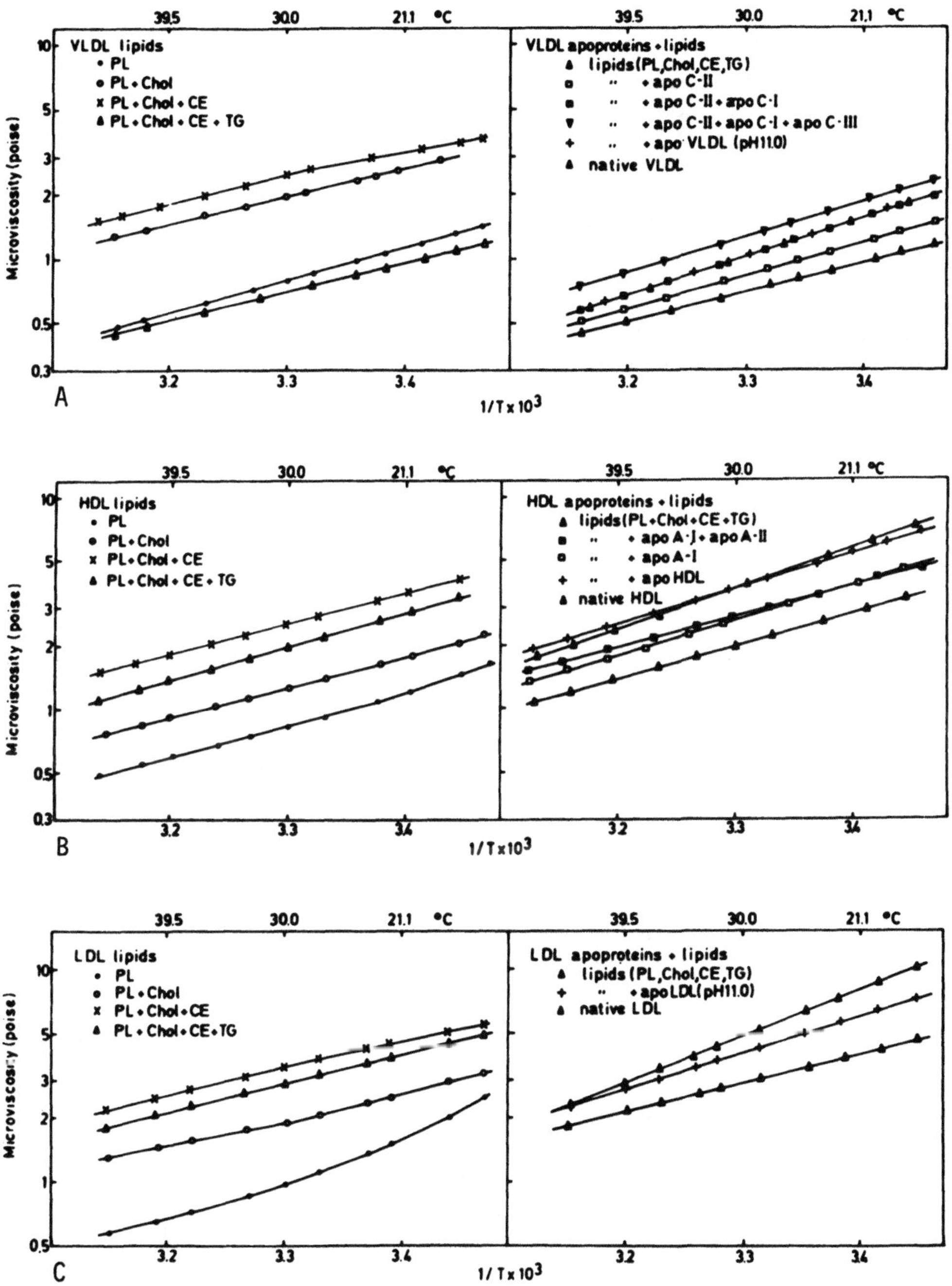

Fig. 6. Microviscosity of stepwise reassembled lipid and lipid plus apoprotein dispersions. A : VLDL; B : HDL; C : LDL.

B. Procedures with incubation

After sonication of the aqueous phospholipid dispersion, the protein can be subsequently incubated with the lipids without sonication at a temperature above or around the transition point. This procedure is specially useful for monitoring either the kinetics of the reaction by spectrophotometry or fluorescence (37), or the heat effect by mixing calorimetry (38). It also avoids any possible denaturation or conformational changes of the protein due to sonic irradiation. Moreover as phospholipid-apoprotein binding are rapid, exothermal processes, the initiation of the reaction does not require any additional energy. This procedure might not be applicable to lipid mixtures containing cholesteryl esters or triglycerides.

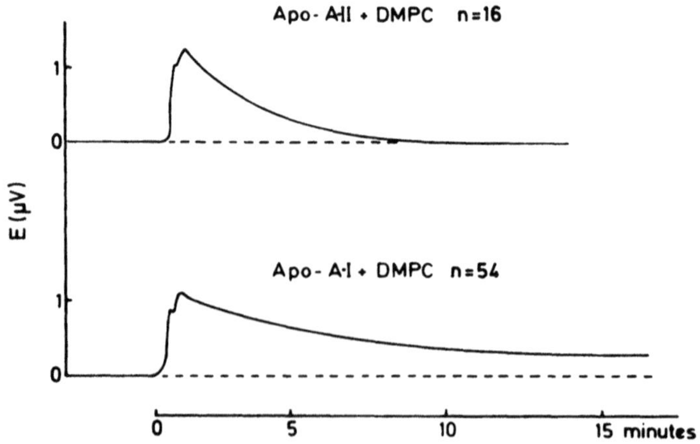

Fig. 7A. Thermograms for the binding of dimyristoyl lecithin liposomes to human apoA-I and apoA-II proteins
n = mole lecithin/mole apoprotein

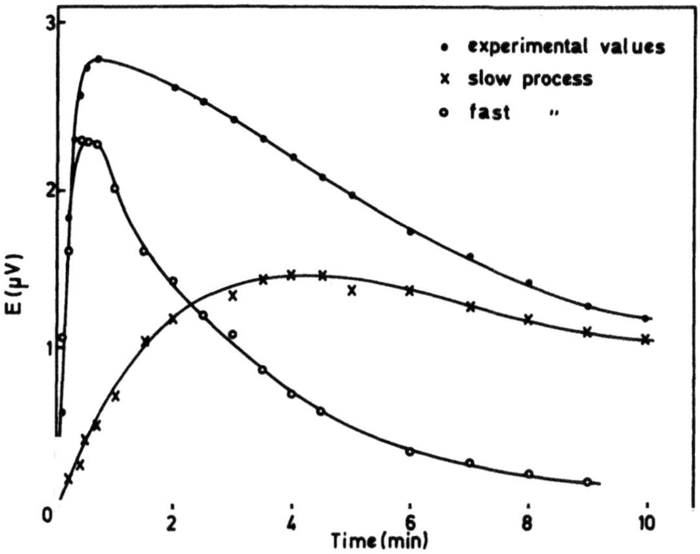

Fig. 7B. Analysis of the above thermogram into two components.

APO A-I - DMPC ASSOCIATION

NATIVE HUMAN APOA-I

N < 100 $\quad$ ApoA-I_P + DMPC $\xrightarrow{\text{FAST}}$ ApoA-I_D $\xrightarrow{\text{FAST}}$ ApoA-I_D-DMPC $\xrightarrow{\text{SLOW}}$ ApoA-I_D-DMPC* + AI_D

POLYMERIC VESICLES $\quad$ DISSOCIATED $\quad$ COMPLEX

N > 100 $\quad$ ApoA-I_P + DMPC $\xrightarrow{\text{FAST}}$ ApoA-I_D $\xrightarrow{\text{FAST}}$ ApoA-I_D-DMPC $\xrightarrow{\text{SLOW}}$ ApoA-I_D-DMPC*

HUMAN APOA-I + 12 MOLE OH-PC

NATIVE BABOON APOA-I

N < 100 $\quad$ ApoA-I_D + DMPC $\xrightarrow{\text{FAST}}$ ApoA-I_D-DMPC $\xrightarrow{\text{SLOW}}$ ApoA-I_D-DMPC*

DISSOCIATED $\quad$ VESICLES $\quad$ COMPLEX

N > 100 $\quad$ ApoA-I_D + DMPC $\xrightarrow{\text{FAST}}$ ApoA-I_D-DMPC $\xrightarrow{\text{SLOW}}$ ApoA-I_D-DMPC*

Fig. 7C. Kinetics of dimyristoyl lecithin vesicles association with native human and baboon apoA-I protein.

IV. MONITORING OF THE REASSEMBLY PROCESS

A. Time progress of the reaction

Fast monitoring of the reaction would involve the use of a stopped-flow spectrophotometer or spectrofluorimeter as described by POWNALL et al (37) for the transfer of pyrene between lipoproteins.

Monitoring over a longer time period can be carried out by various techniques : light scattering detects decrease of the light intensity at 90°, due to the decrease of the size of the complex and to changes in refractive index (39). By electron microscopy a decrease of the size of the multilamellar vesicles and the formation of stacked-discs or "rouleaux" can be visualized (40).

Microcalorimetry does not yield information about the kinetics of fast reactions. However when slow processes are superimposed on the fast association reaction, both effects can be analyzed separately. We applied this technique to the analysis of the apoA-I association with small vesicles of dimyristoyl lecithin (38). On the thermograms depicted in Fig. 7A, the decay curve is slower for apoA-I than for apoA-II-dimyristoyl lecithin (DMPC) association. The thermogram corresponding to apoA-I-DMPC association can be resolved in a fast and slow process by comparison with the experimental curve for apoA-II (Fig. 7B). The fast exothermal process was attributed to dissociation of polymeric apoA-I and binding to DMPC while the slow process would arise from a size decrease and structural reorganization of the complex. This slow process is also temperature dependent and minimal at the phospholipid transition temperature where optimal reassembly occurs (Fig. 7C). Human apoA-I preincubated with lysolecithin and native baboon apoA-I with a low polymer content associate readily with the liposomes.

B. Complex isolation

After completion of the reaction period, the mixture contains free proteins and lipids and a distribution of lipid-protein complexes. Three methods have mainly been applied to the isolation of the various compounds.

1. Sequential ultracentrifugation

This method, developped for the prurification of lipoprotein classes, has also been applied to the isolation of lipid-protein complexes. Free lipids have a density less than 1.06 g/ml, free proteins sediment around 1.23 g/ml, while the complexes have an intermediary density depending on their lipid to protein ratio. Sequential ultracentrifugation involves two steps. Free lipids are first floated at the top at a density of 1.063 g/ml at 200.000 g for 24h. The density of the infranatant is subsequently adjusted at 1.21 g/ml and the complex isolated in the supernatant after a

24h spin at 200.000 g. The protein composition of the complex can be subsequently determined by the method of LOWRY (41) or by amino acid quantitation of an aliquot. The phospholipid content can be derived from a phosphorus analysis according to either BARLETT (42) or EIBL and LANDS (43).

A limitation of the sequential ultracentrifugation technique is the lack of information about the complex density and the possible dissociation of lipid-protein interactions due to 48h spinning in a salt medium.

ASSMANN and BREWER (44) first applied this technique to the analysis of the complexes formed between human apoA-I and apoA-II protein and various phospholipids. We used this method in conjunction with isothermal microcalorimetry to characterize the association of human apo-HDL and apoA-I protein with lysolecithin and dimyristoyl lecithin (45).

2. Gradient ultracentrifugation

A more informative approach to the separation of the components of the reaction mixture is the use of density gradients formed in the preparative ultracentrifuge. Even in this case prolonged ultracentrifugation is required and complex formation might be disturbed by the high concentrations of the gradient components (46). However an advantage of this technique is, that it provides a more detailed picture of the lipid-protein species, assuming that the analysis have been conducted after equilibrium has been reached.

We applied this technique to the isolation and characterization of the complexes formed between human and baboon apoA-I and apoA-II-dimyristoyl lecithin and sphingomyelin. The experimental procedure is as follows : linear gradients made of NaBr spanning the density range 1.06-1.21g/ml were prepared by means of a gradient mixing device and 4.6ml tubes were filled by means of a Büchler Autodensiflo. An approximate amount of 1mg apoprotein + 3mg lipid dispersed in the 1.06g/ml density solution were applied to each tube. Gradient were spun for 66h at 220.000 x g in a SW50 rotor in the Beckman L5-65 centrifuge and subsequently eluted by means of the Autodensiflo.

The optical density at 280nm was recorded by passage through a Pye-Unicam spectrophotometer equipped with a 8 μl cell and 0.3ml fractions were collected. The refractive index of each fraction was measured and the corresponding density calculated to obtain the gradient profiles. Protein and lipids were quantitated in each fraction, according to the methods described above.

Fig. 8A and B illustrate the patterns obtained after association of dimyristoyl lecithin and sphingomyelin with human and baboon apoA-I protein. Complex composition and yields are summarized in Table 1. together with the data obtained on dimeric human apoA-II and monomeric baboon apoA-II proteins (39).

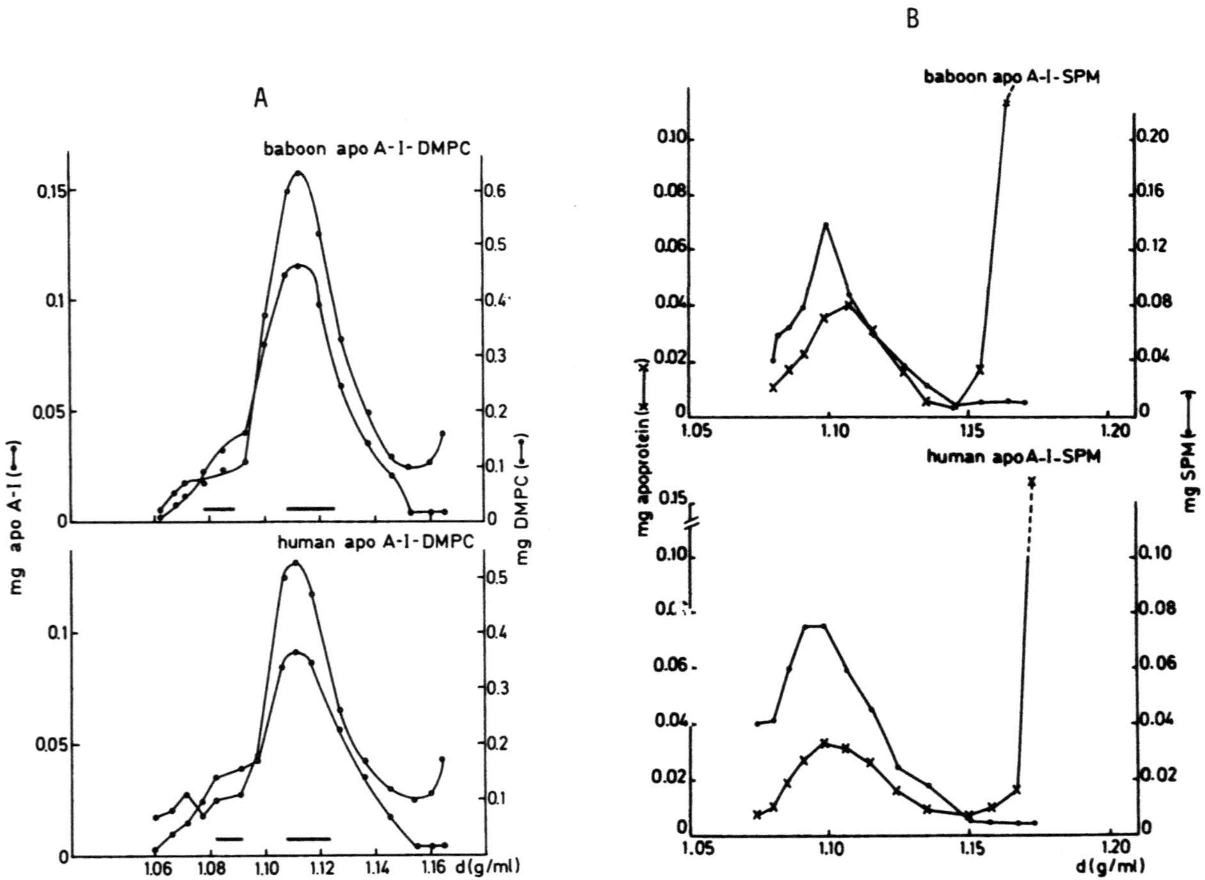

Fig. 8. Gradient ultracentrifugation of human and baboon apoA-I protein and dimyristoyl lecithin (A), sphingomyelin (B).

Table 1

APOPROTEIN	Protein + DMPC d	n	a	b	Protein + SPM d	n	b
ApoA-I (H)	1.09	254.0* ± 12	14	78.4	1.10	84.0 ± 8	49
	1.12	122.0 ± 2	86				
ApoA-I (B)	1.08	248.1 ± 3	20	81.5	1.10	79.0 ± 7	31
	1.12	126.0 ± 3	80				
ApoA-II (H)	1.10	59.8 ± 2	54	74.5	1.09	55.5 ± 5	88
	1.12	45.0 ± 3	46				
ApoA-II (B)	1.10	29.3 ± 1.2	73	82.1	1.09	24.3 ± 3	89
	1.12	22.1 ± 2.0	27				

DMPC : dimyristoyl-L-alpha-phosphatidylcholine
SPM : sphingomyelin
* : mean value of three individual determinations
d : density g/ml

n : $\frac{\text{Mole DMPC}}{\text{Mole apopr.}}$ or $\frac{\text{Mole SPM}}{\text{Mole apopr.}}$

a : amount of complex (%)
b : % apoprotein bound
H : human
B : baboon.

These results suggest that dimyristoyl lecithin associates more easily than sphingomyelin with the apoA-I protein, while the affinity of the apoA-II protein for both phospholipids is similar. The lipid-binding properties of human and baboon apoproteins are very close, as could be expected from the analogies in their amino acid sequence (30). Those data demonstrate that gradient ultracentrifugation offers advantages such as high resolution, good reproducibility and low sample amount. In case of a large lipid excess a first centrifugation step at 1.063g/ml, prior to applying the gradient, avoids any contamination by free lipids.

3. Gel chromatography

Separation of the various components according to their molecular size was performed on Sepharose 6B, 4B and on Biogel AM5 columns equilibrated either in Tris-HCl buffer pH 8.6 or in NH_4HCO_3

0.05M pH 9.0 (47, 44). We isolated the complexes formed between sphingomyelin liposomes and human apoA-I and apoA-II proteins on a Sepharose 6B column (Fig. 9A and B). Free sphingomyelin elute within the void volume while the complexes have a molecular weight of 300.000. 50% free apoA-I protein was still present even though an excess of lipid was present in the reaction mixture.

These results support the data obtained by gradient ultracentrifugation reported above and moreover provide information about the molecular weight of the complex.

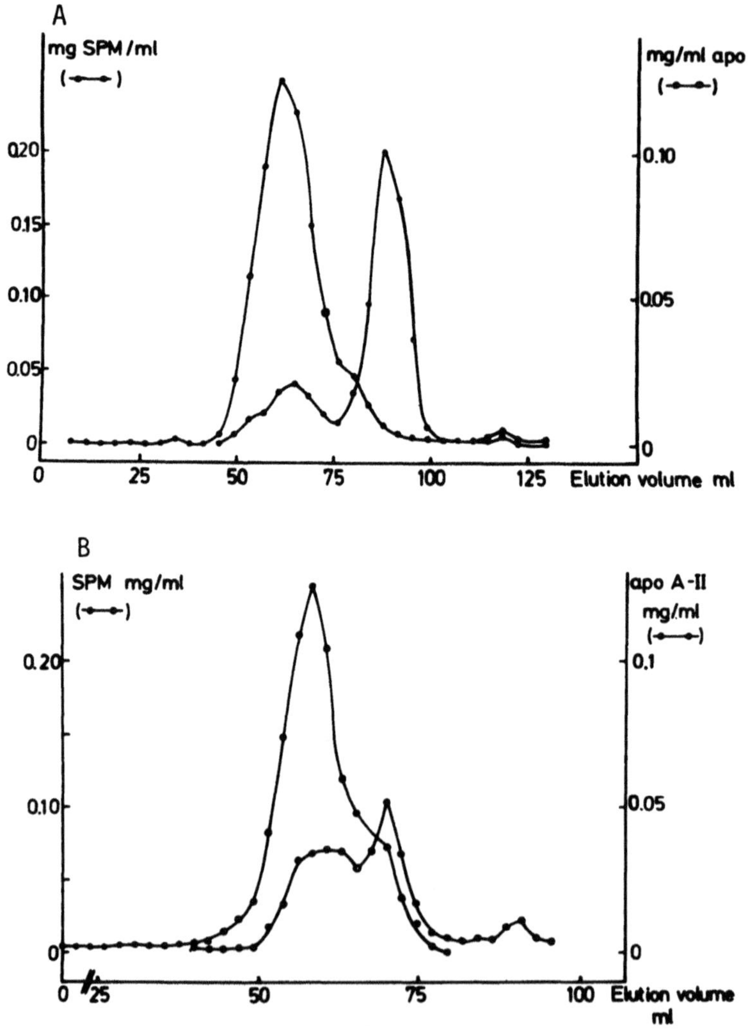

Fig. 9. Isolation of the complexes formed between the sphingomyelin-apoA-I protein (A) and apoA-II protein (B) on Sepharose 6B, 0.05M NH_4HCO_3 pH 9.0

V. CHARACTERIZATION OF THE LIPID-PROTEIN COMPLEXES

A. Compositional analysis

The chemical analysis of the lipid and protein moiety of the reassembled lipoproteins can be performed according to the standard procedures for lipoproteins (35).
A modification of the Lowry procedure was proposed by HERBERT et al (48) to avoid turbidity due to phospholipids i.e. the addition of 0.1ml 0.5M sodium dodecyl sulfate to the reaction mixture.

B. Morphological analysis

By electron microscopy, reassembly mixtures consisting of either apoA-I, apoA-II, apoC-I or apoC-III protein and lecithin appear as stacked discs or "rouleaux" (40). This pattern is changed, by the addition of cholesterol esters, into spheres whose dimensions are similar to those of the native HDL particles (49).

Gel filtration (47), light scattering (39) and sedimentation velocity (33) analysis provide information about molecular weight and size of the reassembled complexes. From fluorescence depolarisation measurement the equivalent volume can be calculated while low-angle X-ray scattering yield the radius of the particle.

C. Structural analysis

The organization of the lipid and protein components within the complex can be derived from the combination of various physical techniques. The secondary structure of the apoprotein is calculated from circular dichroism and optical rotary measurements (50). The calculation of the most probable helical segments of the apoprotein provides information about the nature of the amino acids involved in phospholipid binding (35). The environment of the aromatic amino acids such as tyrosine or tryptophan either in the apoprotein or the complex can be analyzed by differential U.V. spectroscopy or by measurement of the intrinsic tryptophan fluorescence (51).

Potentiometric and calorimetric titration of acidic and basic amino acid residues in the apoA-I, apoA-II, ApoC-I, apoC-III protein and in their complexes with lecithin have demonstrated that a number of charged residues are masked in the complex suggesting that some ionic interactions contribute to the structural stabilization of the complex (52).

The organization of the lipid phase and more specially the degree of freedom the apolar groups can be studied by NMR. ^{13}C, ^{1}H, ^{31}P NMR have provided information about the location and the nature of the hydrophobic interactions within the complex (53). Electron spin resonance, fluorescence depolarization and differential scanning calorimetry have been applied to thermal analysis of the fluidity and phase transition of the lipids. Low angle X-ray scattering yields the election density distribution and locate the polar and apolar components of the complex (54).

D. Thermodynamical analysis

The thermodynamics of lipid-protein interaction are complex as they result from various contributions such as conformational changes of the apoprotein and the phospholipid, transfer from some residues from a polar to a more apolar environment, etc. (45). Though only limited information is available, the change in free energy ΔG can be calculated from the kinetics of a lipid-protein association or from equilibrium dialysis experiments, the enthalpy change ΔH can be measured directly by microcalorimetry (52). The entropy change ΔS can be calculated from the above data and the contribution of ionic and hydrophobic forces evaluated.

VI. INTERACTION OF HDL APOPROTEINS WITH LIPIDS

A. ApoA-I protein

The complexes formed between HDL apoproteins and phospholipids appear as particles stacked in linear arrays with a periodicity of about 50Å when viewed by electron microscopy after negative staining. The individual subunits are discs, with a diameter ranging from 100-200Å. Addition of cholesteryl esters resulted in particles having a morphology similar to that of HDL (40, 49).

By gel filtration and electron microscopy HAUSER et al described the formation of a 200.000 dalton complex between porcine HDL and DMPC (55). We obtained a molecular weight of $2.3.10^5$ daltons for human and baboon apoA-I-DMPC complexes. This complex includes 50% of the apoA-I protein in the reaction mixture (39).

Interaction with sphingomyelin was possible only to a lesser extent (table 1). According to STOFFEL and DÄRR (56) the presence of lysolecithin was necessary in the reaction mixture to obtain high yields of complex formation with a ratio 1:48:13 for apoA-I/lecithin/lysolecithin.

Isothermal microcalorimetry carried out at 28° indicated that apoA-I binds dimyristoyl lecithin exothermally to form a complex which incorporates 80% of the protein in the reaction mixture. The complex composition is 130mole dimyristoyl lecithin/mole apoA-I protein and the enthalpy change is -220Kcal/mole apoA-I. Part of this enthalpy change is due to infolding and increase in helicity of apoA-I under binding (45). This enthalpy change was characterized by an endothermal component at low phospholipid-protein ratios which was attributed to apoA-I dissociation and unfolding and could be reduced by preincubation with lysolecithin. This effect was less pronounced in the baboon apoA-I consisting mainly of dimers instead of dimers and titrations for the human proteins (31).

This incomplete dissociation of apoA-I is an important factor influencing its recombination with phospholipids. Though it can be overcome by reassembly with small unilemellar DMPC vesicles, it impedes reassembly with egg-lecithin and sphingomyelin heterogenous vesicles. This might account for the literature discrepancies concerning phospholipid binding to this apoprotein.

We measured the microviscosity of dimyristoyl lecithin liposomes after labelling with 1.3 diphenyl, 1, 3, 5 hexatriene (DPH) by following the degree of fluorescence polarization as a function of temperature (Fig. 10.).

In the presence of increasing quantities of apoA-I protein the phospholipid transition occuring at 24° disappears gradually, while a new transition appears shifting to higher temperatures under addition of apoA-I (Table 2).

The presence of apoA-I fluidizes the acyl chains at low temperatures and increases η at t > transition temperature.

B. ApoA-II protein

The second major component of apo-HDL has been characterized by several authors with respect to its sequence, self-association properties and lipid-binding characteristics (50). This apoprotein occurs either as a dimer in man or chimpanzee or a monomer in rhe-

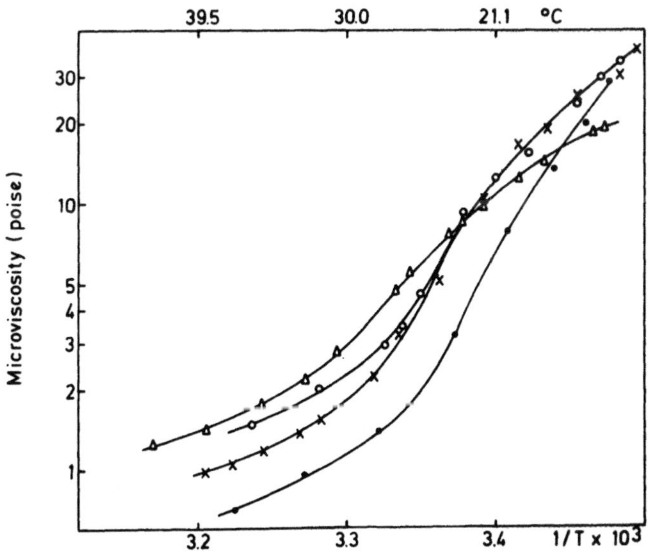

Fig. 10. Microviscosity of dimyristoyl lecithin apoA-I protein mixtures () DMPC; (x) n = 120; (o) n = 60; (Δ) n = 17 $\frac{\text{mole DMPC}}{\text{mole apoA-I}}$

Table 2 Phase transition parameters of DMPC/apoprotein complexes

Complexes	Microviscosity (poise) 15°C	40°C	Low Temp. Transition T_{M1}	Range	ΔT	High Temp. Transition T_{M2}	Range	ΔT
DMPC	18.00	0.63	24.0	21.9 - 26.1	4.2			
Mol DMPC / Mol apoA-I								
N=120	24.29	0.91	23.5	21.8 - 25.2	3.4	25.5	23.7 - 28.4	4.7
N= 60	24.29	1.13				25.5	23.6 - 28.3	4.7
N= 17	15.04	1.29				27.5	25.6 - 33.0	7.4
Mol DMPC / Mol apoA-II								
N= 70	26.75	1.08	24.0	21.9 - 27.3	5.4	27.5	23.3 - 31.4	8.1
N= 35	22.21	1.26				28.0	24.0 - 32.4	8.4
N= 16	10.43	1.38				29.5	24.8 - 32.4	7.6
Mol DMPC / Mol apoC-I								
N=125	25.06	0.79	23.5	21.0 - 26.0	5.0	25.5	22.5 - 29.0	6.5
N= 62	26.00	0.89				25.5	21.8 - 29.2	7.4
N= 21	25.06	1.12				26.5	22.7 - 30.2	7.5
Mol DMPC / Mol apoC-III								
N=120	32.80	0.70	24.0	21.8 - 26.2	4.4	26.5	24.2 - 31.2	7.0
N= 65	26.75	0.74	24.0	21.9 - 26.1	4.2	27.0	22.8 - 31.2	8.4
N= 21	32.07	0.84	26.0	22.7 - 29.6	6.9	28.5	25.4 - 31.6	6.2

sus and baboon. The presence of the cysteine does not affect its affinity for phospholipids (30). According to LUX et al (1), ASSMANN and BREWER (44), the lipid-binding segments in human apoA-II protein are located at the carboxyl-terminal end of the protein. Our potentiometric data support this interpretation and suggest that 6 out of 9 lysines are masked by phospholipids in the apoA-II lecithin complex (57, 58).

We have compared the affinity of human and baboon apoA-II protein for lecithin and sphingomyelin and obtained similar data for both phospholipids, whereas for apoA-I protein a preferential binding to lecithin was observed (Table 1). Moreover the kinetics of the apoA-II-phospholipid association are faster than for apoA-I protein and the reaction more exothermal (Fig. 7A). This apoprotein seems therefore to bind lipids more readily than apoA-I protein. The shift of the transition temperature of dimyristoyl lecithin is more significant upon addition of apoA-II than apoA-I protein. The same was observed for the fluidization of the acyl chains at t >

transition temperature and their immobilization at t < transition temperature (Fig. 11.).

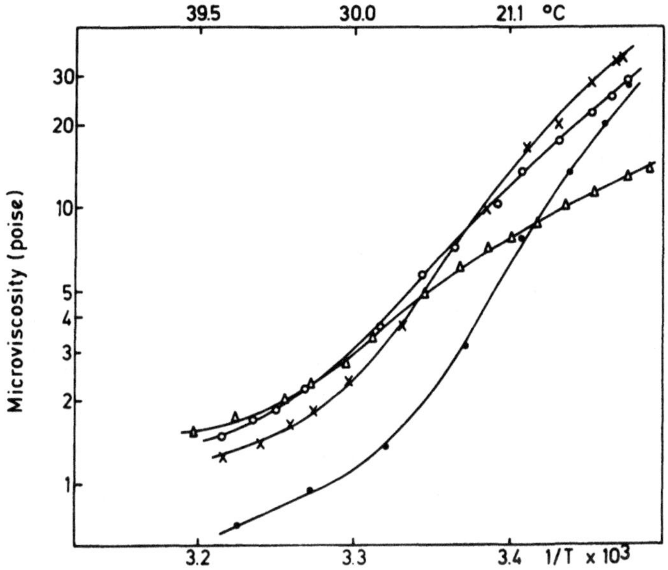

Fig. 11. Microviscosity of DMPC-apoA-II mixtures (•) DMPC; (x) n = 70; (o) n = 35; (Δ) n=16

$$n = \frac{\text{mole DMPC}}{\text{mole apoA-II protein}}$$

VII. INTERACTION OF VLDL APOPROTEINS WITH LIPIDS

The interaction of apoC-I, apoC-II, apoC-III proteins with lipids is well documented and presents a number of characteristics in common (59, 60).

An increase in helical content as well as a shift in the tryptophan fluorescence maximum was observed, the complex composition was 60 moles lecithin/mole apoC-III protein. We measured the enthalpy change on binding dimyristoyl lecithin to apoC-I and apoC-III protein (61) and observed the changes in microviscosity during complex formation.

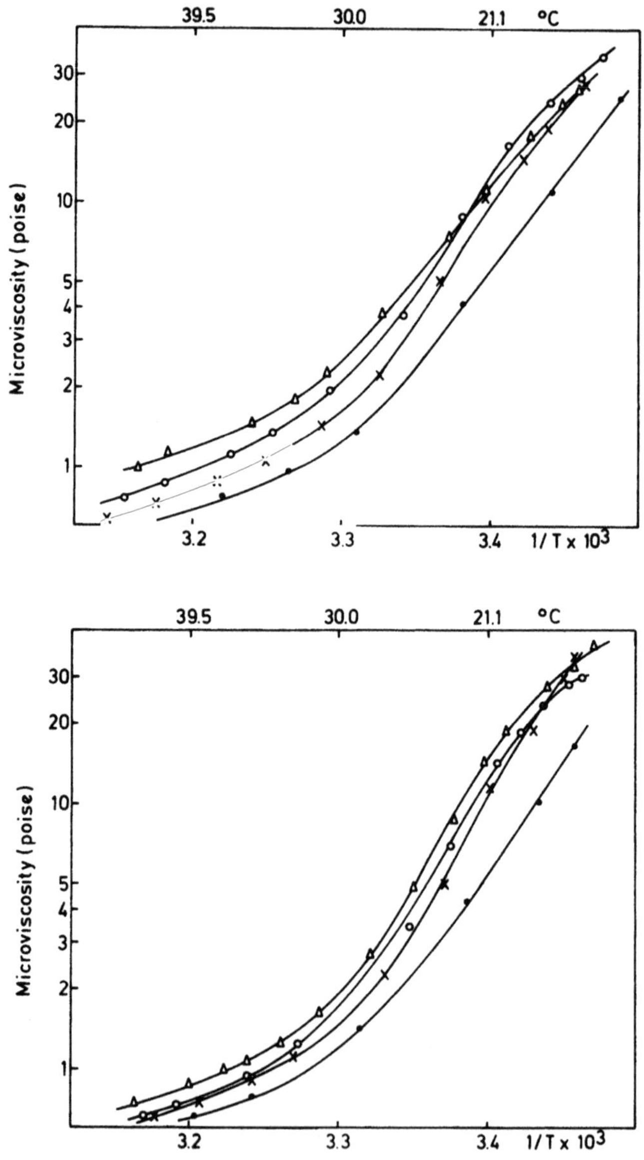

Fig. 12. Microviscosity of apoC-I/DMPC (A) and apoC-III/DMPC (B) mixtures
A (•) DMPC; (x) n = 125; (o) n = 62; (Δ) n = 21mole DMPC/apoC-I

B (•) DMPC; (x) n = 120; (o) n =65; (Δ) n = 21mole DMPC/mole apoC-III

By comparison of the titration data of the native apoproteins and of their complexes with lecithin, location and structure of lipid-binding segments was proposed. These results agree with those of relipidation studies on native or synthetic apoC-I and apoC-III fragments (58).

CONCLUSIONS

The reassembly of apoproteins and lipids involves the knowledge of the nature of the constituents, of their physical properties in solution and the analysis of their behaviour in the reassembled state. Thanks to the use of model systems, the information in this area becomes more accurate and more relevant to the physiological function. An ultimate test on the validity of any "in vitro" reassembly experiment is of course the analysis of the biological activity of the system and its aptitude to simulate the "in vivo" situation. Through this bias the use of such systems either semi or completely synthetic might prove valuable to replace deficient functioning entities for lipid transport and turnover during genesis and development of atherosclerosis.

ACKNOWLEDGEMENTS

We are endebted to Mrs.J.LIEVENS, Mr.R.VERCAEMST and Mr.H.CASTER for skilful technical assistance. This research was supported by Grant 3001075 NFGWO, Brussels.

REFERENCES

1. S.E.LUX, R.HIRZ, R.I.SHRAGER, A.M.GOTTO
J.Biol.Chem. 247, 2598 (1972).

2. A.M.SCANU, C.T.LIM, C.T.EDELSTEIN
J.Biol.Chem. 247, 5850 (1972).

3. A.M.SCANU, E.CUMP, J.TOTH, S.KOGA, E.STILLER, J.ALBERS
Biochemistry 2, 1327 (1970).

4. R.B.HAMILTON, H.J. KAYDEN
"Biochemistry of Disease", 5, 531, F.F.Becker, ed., Dekker, New York (1974).

5. C.TANFORD
"The Hydrophobic Effect" p.121
Wiley, New York (1973).

6. R.B.VERDERY, A.NICHOLS
Chem.Phys.Lipids 14, 123 (1975).

7. M.ROSSENEU, F.SOETEWEY, V.BLATON, J.LIEVENS, H.PEETERS
Chem.Phys.Lipids 13, 203 (1974).

8. H.HAUSER, M.C.PHILLIPS
J.Biol.Chem. 248, 8585 (1973).

9. A.N.KRUSKI, A.M.SCANU
Chem.Phys.Lipids 13, 27 (1974).

10. V.LUZZATI
in "Biological Membranes" p.71
ed. D.Chapman, Acad.Press, New York (1968).

11. R.M.WILLIAMS, D.CHAPMAN
Prog. Chem. Fats and other Lipids 11, 1 (1970).

12. D.CHAPMAN, D.F.H.WALLACH
in "Biological Membranes" p.125,
D.Chapman ed., Acad.Press, New York (1968).

13. A.D.BANGHAM, R.W.HORNE
J.Mol.Biology 8, 1660 (1964).

14. K.J.PALMER, F.O.SCHMITT
J.Cell.Comp.Physiol. 17, 385 (1941).

15. A.D.BANGHAM, M.M.STANDISH, J.C.WATKINS
J.Mol.Biol. 3, 238 (1965).

16. C.H.HUANG
Biochemistry 8, 344 (1969).

17. J.SURKUUSK, B.R.LENTZ, Y.BARENHOLZ, R.L.BILTONEN, T.E.THOMPSON
Biochemistry 15, 1393 (1976).

18. M.C.PHILLIPS, H.HAUSER, R.B.LESLIE, D.OLDANI
Biochim.Biophys.Acta 406, 402 (1975).

19. S.BATZRI, E.D.KORN
Biochim.Biophys.Acta 298, 1015 (1973).

20. J.BRUNNER, P.SKRABAL, H.HAUSER
Biochim.Biophys.Acta 455, 322 (1976).

21. B.LUNDBERG
Chem.Phys.Lipids 18, 212 (1977).

22. B.LUNDBERG, E.R.SAARINEN
Chem.Phys.Lipids 14, 260 (1975).

23. B.SHORE, V.SHORE
Biochemistry 8, 4510 (1969).

24. W.V.BROWN, R.I.LEVY, D.S.FREDRIKSON
J.Biol.Chem. 245, 6588 (1970).

25. V.BLATON, R.VERCAEMST, N.VANDECASTEELE, H.CASTER, H.PEETERS
Biochemistry 13, 1127 (1974).

26. J.C.OSBORNE, G.PALUMBO, B.BREWER
Biochemistry 15, 317 (1976).

27. J.GWYNNE, H.B.BREWER, H.EDELHOCH
J.Biol.Chem. 250, 2269 (1975).

28. L.B.VITELLO, A.M.SCANU
J.Biol.Chem. 251, 1131 (1976).

29. L.B.VITELLO, A.M.SCANU
Biochemistry 15, 1161 (1976).

30. V.BLATON, R.VERCAEMST, M.ROSSENEU, J.MORTELMANS, R.L.JACKSON, A.M.GOTTO, H.PEETERS
Biochemistry 16, 2157 (1977).

31. M.ROSSENEU, V.BLATON, R.VERCAEMST, F.SOETEWEY, H.PEETERS
Eur.J.Biochem. 74, 83 (1977).

32. A.JONAS
Biochim.Biophys.Acta 393, 471 (1975).

33. W.L.STONES, J.A.REYNOLDS
J.Biol.Chem. 250, 8045 (1975).

34. A.M.SCANU
J.Lipid.Res. 7, 295 (1966).

35. A.M.SCANU, C.EDELSTEIN, P.KEIM
in "The Plasma Proteins : Structure, Function and Genetic Control",
F.Putnam ed. Acad.Press, New York, p.317 (1975).

36. M.ROSSENEU, F.SOETEWEY, R.VERCAEMST, J.LIEVENS, H.PEETERS
Prot.Biol.Fluids 25, 47 (1977).

37. H.J.POWNALL, J.B.MASSEY, J.T.SPARROW, A.M.GOTTO
Prot.Biol.Fluids 25, 115 (1977).

38. M.ROSSENEU, F.SOETEWEY, J.LIEVENS, H.PEETERS
Prot.Biol.Fluids 25, 121 (1977).

39. M.ROSSENEU
Aggregaatsthesis, Univ. of Antwerpen (1977).

40. T.M.FORTE, A.V.NICHOLS, E.L.GONG, S.LUX, R.I.LEVY
Biochim.Biophys.Acta 248, 381 (1971).

41. O.H.LOWRY, H.J.ROSEBROUGH, A.L.FARR, R.J.RANDALL
J.Biol.Chem. 193, 265 (1951).

42. G.R.BARTLETT
J.Biol.Chem. 234, 466 (1959).

43. H.EIBL, W.E.LANDS
Biochemistry 9, 423 (1970).

44. G.ASSMAN, H.B.BREWER
Proc.Nat.Acad.Sci.US, 47, 1309 (1974).

45. M.ROSSENEU, F.SOETEWEY, G.MIDDELHOFF, H.PEETERS, W.V.BROWN
Biochim.Biophys.Acta 441, 68 (1976).

46. A.M.SCANU, J.L.GRANDA
Biochemistry, 5, 446 (1966).

47. G.MIDDELHOFF, M.ROSSENEU, H.PEETERS, W.V.BROWN
Biochim.Biophys.Acta 441, 57 (1976).

48. P.N.HERBERT
Personal Communication.

49. T.FORTE, E.GONG, A.V.NICHOLS
Biochim.Biophys.Acta 551, 169 (1974).

50. J.D.MORRISETT, R.L.JACKSON, A.M.GOTTO
Ann.Rev.Biochem. 44, 183 (1975).

51. C.J.HART, R.B.LESLIE
Chem.Phys.Lipids 4, 369 (1970).

52. H.PEETERS, M.ROSSENEU
in "Living Systems as Energy Converters"
R.Buvet, J.P.Massué ed., Elsevier Publ. p.55(1977).

53. B.SEARS, R.J.DECKELBAUM, M.J.JANIAK, G.G.SHIPLEY, D.M.SMALL
Biochemistry 15, 4151 (1976).

54. G.G.SHIPLEY, D.ATKINSON, A.M.SCANU
J.Supramol.Struct. 1, 98 (1972).

55. H.HAUSER, R.HENRY, R.B.LESLIE, J.M.STUBBS
Eur.J.Biochem. 48, 583 (1974).

56. W.STOFFEL, W.DÄRR
Hoppe-Zeyler's Z. Physiol.Chem. 357, 127 (1976).

57. M.ROSSENEU, F.SOETEWEY, M.J.LIEVENS, R.VERCAEMST, H.PEETERS
Eur.J.Biochem. 79, 251 (1977).

58. F.SOETEWEY, M.J.LIEVENS, R.VERCAEMST, M.ROSSENEU, H.PEETERS, V.W. BROWN
Eur.J.Biochem., 79, 259 (1977).

59. R.L.JACKSON, J.D.MORRISETT, R.T.SPARROW, J.P.SEGREST, H.J.POWNALL, L.C.SMITH, H.F.HOFF, A.M.GOTTO
J.Biol.Chem. 249, 5314 (1974).

60. J.D.MORRISETT, J.S.K.DAVID, H.J.POWNALL, A.M.GOTTO
Biochemistry 12, 1290 (1973)

61. M.ROSSENEU, F.SOETEWEY, H.PEETERS, L.BAUSSERMAN, P.N.HERBERT
Eur.J.Biochem., 70, 285 (1976).

Metabolism of the Lipoprotein Molecule

INTERCONVERSION OF PLASMA LIPOPROTEINS

Alex V. Nichols

Donner Laboratory, Lawrence Berkeley Laboratory

University of California, Berkeley, CA. 94720 USA

INTRODUCTION

Interconversion of a lipoprotein molecule to a lipoprotein molecule of different physical and chemical properties perhaps can be best considered in terms of our current understanding of lipoprotein structure. Thus, normal lipoproteins are generally described as spherical structures with a core of triglycerides and cholesteryl esters stabilized by a surface of specific apoproteins, phospholipids and cholesterol (1, 2). On the surface, the phospholipids and cholesterol form a monolayer in which apoproteins exhibiting amphipathic helical regions are embedded. This overview will consider interconversions which result from significant alterations in molecular size and/or shape due to enzymatically or physicochemically-induced changes in core and/or surface domains of a lipoprotein molecule. Analytic approaches to the investigation of interconversions include: preparative and analytic ultracentrifugation, electron microscopy, gel chromatography and electrophoresis, and chemical analysis.

INTERCONVERIONS: ENZYME-INDUCED

Lipoprotein Lipase: Interconversion of Very Low Density Lipoproteins

Hydrolytic depletion of core triglycerides in very low density lipoproteins (VLDL) by post-heparin lipolytic activity (PHLA) produces lipoprotein species of smaller size and higher particle density (3). The chemical properties of these species indicate a process of core reduction accompanied by decrease of surface material in excess of that required to stabilize the remaining triglyceride core. Transfer of C apoproteins and phospholipid from the VLDL surface to high density lipoproteins (HDL) during removal of VLDL triglyceride from the circulation has been described (4) and would appear to be associated with the process of surface adjustment during hydrolytic core reduction. In certain states of abnormal lipid metabolism, accumulations of lipoprotein surface lipids are observed in vivo in the form of abnormal particles in the plasma. Thus, in patients deficient in plasma lecithin:cholesterol acyltransferase (LCAT), abnormal vesicular species rich in phospholipid and cholesterol appear in the d 1.006-1.019 and d 1.019-1.063 fractions and are considered to arise during degradation of chylomicrons and VLDL by lipoprotein lipase (LPL) in vivo (6). Such vesicular species have not been observed in plasma of normal subjects and it would appear that disposal of surface components is efficiently handled under normal conditions. Enzymatic mechanisms proposed for disposition of lipoprotein surface lipids include: (1) hydrolysis of the phospholipids by phospholipase activity associated with LPL, and (2) conversion by LCAT of surface phosphatidylcholine and unesterified cholesterol to lysolecithin and cholesteryl esters (7). The interconversions noted in triglyceride-rich lipoproteins during PHLA activity are considered similar to those occurring during normal metabolic clearance of these species, where "remnants" of chylomicrons and VLDL are formed during hydrolytic encounters with LPL situated on capillary endothelial surfaces (8). An interesting aspect of the above interconversion pathway is the apparent persistence of association of one major apoprotein (apoB) with the residual core lipid leading to the eventual formation of the low density lipoprotein (LDL) class (9). Differential affinities of other apoproteins for the lipid moiety remaining during the course of VLDL interconversion are also suggested by the differing apoprotein compositions of VLDL subspecies of increasing hydrated density (10). One of the released apoproteins is the activating cofactor for LPL (apoC-II). Precursors to VLDL particles secreted by the liver are also spherical particles but of somewhat different chemical composition particularly at the surface (11, 12). These particles pick up C apoproteins from HDL in blood and with some additional transfers of surface lipid assume the composition of VLDL usually encountered in blood (13). Apparently, no other major structural transformations of precursor plasma chylomicrons and VLDL occur until the particles encounter the hydrolytic activity of LPL.

Lecithin:Cholesterol Acyltransferase:Interconversion of Discoidal to Spherical HDL

Another major interconversion process occurring in vitro and probably in vivo involves the enzymatic build-up rather than degradation of an apolar lipid core in a lipoprotein molecule. This interconversion has been designated the "disc-to-sphere" interconversion wherein discoidal lipoprotein complexes, composed primarily of phospholipid, unesterified cholesterol and specific apoproteins and deficient in core lipids, are transformed by LCAT to spherical structures during build-up of an apolar core of cholesteryl ester (14, 15, 16). The spherical particles formed have properties similar to normal HDL. Discoidal complexes detected in the d 1.063-1.20 fraction of plasma of human subjects with LCAT-deficiency and cholestasis transform to spherical HDL particles when plasma of such subjects is incubated with LCAT (14). The similarity of the spherical products to normal HDL and the occurrence of discoidal particles in plasma of certain human patients have prompted the proposal that such discoidal complexes may be precursor HDL which are converted to normal HDL after secretion and interaction with LCAT.

Rat liver perfusion studies. Animal data supporting the idea that discoidal complexes are precursor forms of HDL (nascent HDL), secreted by the liver and converted to spherical HDL by LCAT activity, have recently been reported by Hamilton et al. (17). Using the perfused rat liver, these workers demonstrated discoidal complexes (diameter: 190 ± 25 Å; thickness: 46 ± 5 Å) in a perfusate containing an LCAT inhibitor (5,5′-dithionitrobenzoic acid, DTNB). In the presence of DTNB the perfusate contained predominantly (∿ 75% of particles observed) spherical species (diameter: 75-125 Å) which were similar to normal rat HDL (diameter: 114 ± 13 Å). The discoidal complexes were rich in surface lipids (phospholipids and unesterified cholesterol) and deficient in core lipids. The HDL fraction from the noninhibited perfusate (containing spherical particles and some discs) showed a level of core lipid approaching, but not equal, to that of normal rat HDL. The apoprotein composition by polyacrylamide gel electrophoresis was markedly different in the discoidal and spherical HDL preparations. Hence, while apoE (arginine-rich apoprotein) was found to be the major component of the discoidal species from the inhibited perfusate, apoA-I was the primary component of the HDL from the noninhibited perfusate rich in spherical particles. The above would suggest that during the LCAT-induced disc-to-sphere interconversion, apoA-I is incorporated and apoE is to some extent dissociated. However, an alternative explanation might be that ultracentrifugation in high salt dissociates apoA-I from discoidal complexes and apoE from spherical HDL. Indeed, results consistent with dissociation of some apoE from rat plasma spherical HDL (rich in apoA-I) during ultracentrifugation have been reported (18, 19). Although dissociation of apoA-I from rat discoidal complexes during ultracentrifugation has not been evaluated, possible dissociation of apoA-I from lipoproteins rich in surface lipids (i.e., discs) has been noted during ultracentrifugation of LCAT-deficient plasma (20). Thus, following preparative ultracentrifugation of LCAT-deficient plasma, there is considerable apoA-I in the d > 1.25 fraction. Interestingly, a substantial amount of this apoA-I is incorporated into the d 1.063-1.20 fraction following incubation of the plasma with LCAT, suggesting that under conditions of ultracentrifugation, apoA-I may have a greater affinity for cholesteryl ester-rich particles than for surface lipid-rich particles (20). Hence, in terms of the available data, it is not possible to establish whether the observed apoprotein compositional differences reflect an interconversion process whereby apoA-I (from some unidentified source) is incorporated and some apoE is dissociated during formation of a cholesteryl ester core in discoidal particles; or whether both apoproteins are initially present in the discs and spheres but are dissociated during ultracentrifugation in accord with their differential affinities for lipid mixtures deficient or enriched in core lipid. At present no data are available on the physical-chemical changes occurring in discoidal complexes during in vitro incubation with LCAT of a rat liver perfusate collected under conditions where LCAT was inhibited. Hence, no comparison is possible between spherical products produced utilizing perfusate discoidal complexes as substrate and those appearing in plasma or in the noninhibited perfusate of rat liver where the conversion is presumed to occur upon secretion of discoidal complexes into an LCAT-containing medium. Clearly the sequence of events whereby the final apoprotein composition of normal rat plasma HDL is attained, during the disc-to-sphere interconversion, requires further investigation.

Additional definition of the molecular species involved in the interconversion process is probably also needed. For example, electrophoresis of the HDL from the rat liver perfusate, in which LCAT activity is inhibited, indicates the presence of two lipoprotein species, whereas predominantly one (discoidal complexes) is observed under electron microscopy (17). Data from studies on LCAT-deficient plasma have demonstrated the participation of HDL species (small molecular weight HDL particles with diameters of 50-60 Å) other than discoidal complexes in the interconversion process (21). Thus, the interconversion process may involve not only direct transformation of discs to spheres, but also participation of some intermediate subspecies which could act as alternate sites for cholesteryl ester build-up and formation of particles similar to normal HDL.

Lastly, in view of the apparent ability of LCAT to transform diverse substrate particles to relatively homogeneous HDL species, it is possible that the precise form of the substrate is not crucial and that an LCAT-apoA-I complex is the critical HDL-generating unit which functions by organizing apoprotein and polar lipid moieties around a growing cholesteryl ester core. Inherent lipid binding affinities and capacities of specific apoproteins would determine the final lipoprotein composition. The above discussion indicates the need for detailed studies on purified systems where interconversions of well-defined model substrates can be followed and characterized by physical and chemical means.

Incubation of plasma of LCAT-deficient subjects with LCAT. The properties of the HDL particle distribution in plasma of LCAT-deficient patients, which includes discoidal and spherical small molecular weight HDL particles, have been rather extensively characterized. Some major differences exist between HDL of LCAT-inhibited rat liver perfusate and HDL of LCAT-deficiency. Unlike the discoidal HDL in the rat liver perfusate, the discoidal HDL in LCAT-deficient plasma has a significant content of both apoA-I and apoA-II (22). Interestingly, however, the disc-containing fractions from both sources have a substantial content of apoE (20, 23). The small molecular weight HDL particles in LCAT-deficient plasma also contain substantial apoA-I and apoA-II. As noted previously, apoA-I is found not only in the d 1.063-1.20 fraction, but also in the d > 1.25 bottom fraction. This suggests that some apoA-I either exists freely in such plasma or is loosely bound to phospholipid-cholesterol-rich lipoprotein species and is dissociated upon ultracentrifugation in high salt. As previously noted, in the presence of LCAT activity, some of this apoA-I is incorporated into d 1.063-1.20 particles resembling HDL. It would appear that formation of apolar core material stabilizes the binding of apoA-I to lipid, perhaps by promoting formation of additional amphipathic helical conformation in the apoprotein. It is also of interest to note that, during incubation of LCAT-deficient plasma with LCAT, the content of apoE is markedly reduced in the HDL fraction, with the released apoE now appearing primarily in the VLDL fraction (20). A detailed investigation of the disc-to-sphere interconversion occurring during incubation of LCAT-deficient plasma with partially-purified LCAT has been reported (20). In this study, the interconversion was followed at two levels of added LCAT activity. At a low level of LCAT activity due to presence of p-chloromercuriphenyl-sulfonate [PCMPS] (LCAT activity approximately one-tenth of that in presence of mercaptoethanol [ME]), there was a partial reduction in content of discoidal species (diameter: 170-240 Å) which form the large molecular weight end of their HDL distribution. This reduction was associated with the appearance of spherical HDL species (mean diameter: ∿ 78 Å) smaller than the discs but somewhat larger than the small molecular weight HDL (diameter: 50-60 Å) usually encountered in LCAT-deficient plasma. At full LCAT activity (in presence of ME) there was further reduction in content of discoidal complexes (almost full disappearance by electron microscopy); and a corresponding increase in smaller molecular weight species which now exhibited a still larger average molecular weight and size (mean diameter: ∿ 98 Å) than either the usual small molecular weight species or the small molecular weight product observed at low level LCAT activity. These data suggest that LCAT, utilizing a most heterogeneous mixture of possible substrate material in LCAT-deficient plasma (discs, small molecular weight HDL, loosely-bound apoA-I, etc.), initially forms small spherical molecular species (diameter ∿ 78 Å) which, with further LCAT activity, are transformed to still larger spherical species (diameter ∿ 98 Å) within the HDL par-

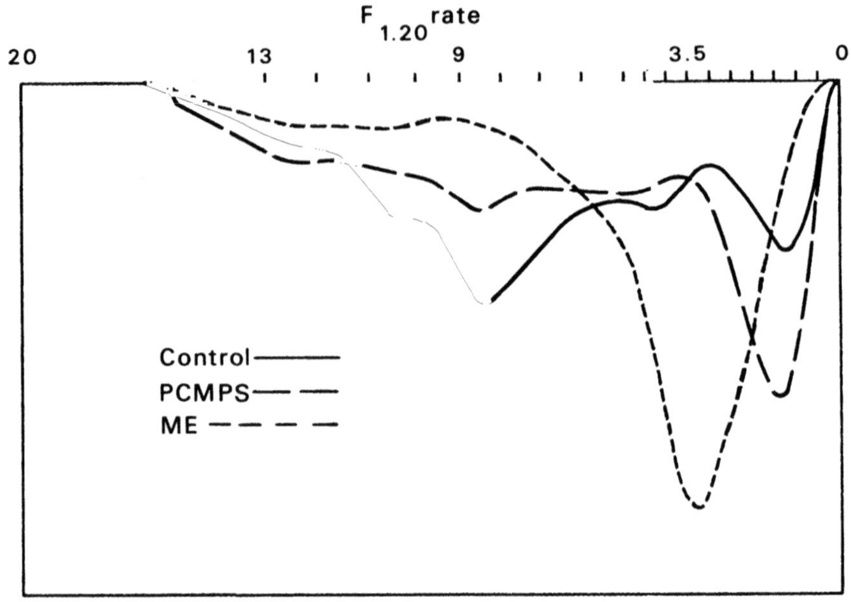

Figure 1: Analytic ultracentrifugal patterns showing the effect of incubation (24 hr, 37°C) with partially purified LCAT on the HDL in plasma of LCAT-deficient patient (M.R.) in presence of PCMPS (2 mM) and ME (10 mM).

ticle distribution. This interconversion is perhaps best appreciated by reference to the analytic ultracentrifugal patterns obtained for the above incubation mixtures (Fig. 1). In these patterns the discoidal species exhibit a peak whose flotation rate is $F^{\circ}_{1.20}$ 8.4, while the peak of the small molecular weight spherical particles exhibits a flotation rate of $F^{\circ}_{1.20}$ 1.25. The control pattern shows the presence of both of the above species. At low level LCAT activity, the build-up of smaller molecular weight species shows a peak at $F^{\circ}_{1.20}$ 1.4, while at full LCAT activity the further build-up shows a peak primarily at $F^{\circ}_{1.20}$ 3.3.

In view of the rather close correspondence of the ultracentrifugal and chemical properties of the above products to those of recently identified major components (24) within the normal HDL distribution, these observations imply that such interconversion processes may be involved in the origin of the normal HDL distribution. Thus, recent work by Anderson et al. (24) has established the existence of at least three major components within the HDL particle distribution from normal human plasma. These components have the following hydrated densities and mean peak flotation rates: HDL_{2b}, d 1.063-1.100, $F^{\circ}_{1.20}$ 5.37; HDL_{2a}, d 1.100-1.125, $F^{\circ}_{1.20}$ 3.15; and HDL_3, d 1.125-1.200, $F^{\circ}_{1.20}$ 1.56. Comparison of the flotation rates and chemical compositions of these three components with those of the two small molecular incubation products (obtained at low and full level LCAT activity) shows rough similarity between the low level activity product and HDL_3, and the full activity product and HDL_{2a}. Complete comparison, however, will require information on the apoprotein compositions of both the incubation products and the HDL components. Nevertheless, the above incubation data do suggest a possible sequence in the origin of the HDL particle distribution, wherein LCAT activity first organizes the known substrate and cofactor components into species resembling HDL_3, and, with the availability of sufficient substrate components and further activity, into additional species resembling HDL_{2a}.

In any consideration of such a step-wise build-up of the HDL distribution it is of interest to note the steady state distribution of the plasma total HDL concentration among the three HDL components in a population sample of normal human subjects. The results of such a three component analysis show that the distribution of total HDL concentration among the three components is significantly correlated with the total HDL concentration (25). Thus, in cases with total HDL concentrations in the range of 100 ± 5 mg/dl, over 90% of the total HDL concentration is represented by HDL_3. At higher total HDL levels (100-200 mg/dl), the increase in concentration is contributed primarily by the HDL_{2a} component with little change in HDL_3. At still higher concentrations of total HDL (200-475 mg/dl), there is further increase in HDL_{2a} levels accompanied now by the appearance and build-up of HDL_{2b} levels. At very high total HDL concentrations (> 475 mg/dl), levels of HDL_{2a} and HDL_{2b} are further increased in conjunction with the appearance of additional faster floating HDL components. Thus, the average trends in distribution of the total HDL distribution among the three components suggest processes, possibly involving LCAT, which distribute increasing plasma levels of HDL lipid substrates and apoproteins in a directed manner into lipoprotein species of larger particle size and lower hydrated density. It will be of considerable interest to establish whether such interrelationships between plasma levels of these three lipoprotein components hold in individuals during short-term reduction or elevation of their total HDL levels. Although the above considerations are for the most part speculative, they do point to some exciting and challenging problems in the structure and function of the HDL particle distribution.

INTERCONVERSIONS: RECENT PHYSICAL-CHEMICAL STUDIES

As noted earlier, lipoprotein interconversions may result from alterations either in the surface or the core domains of a lipoprotein molecule. Of considerable interest are the effects of alterations restricted predominantly to one domain, which are not complicated by concomitant major alterations in the other. Recently, two very dissimilar approaches have been applied to the HDL molecule, which dissociate some of its surface apoprotein and produce interconversions of physical-chemical interest and of potential biological significance. The first approach utilizes the protein denaturant, guanidine-HCl (GuHCl) (26) while the second uses a nonphysiologic phospholipid, dimyristoylphosphatidylcholine (DMPC) (27, 28).

Interconversions of HDL Induced by Guanidine-HCl

Perturbation of HDL structure by a sequence of GuHCl exposure and removal is characterized by: (1) selective dissociation of lipid-poor apoA-I, which follows a time-course similar to denaturation-related changes in spectroscopic parameters (e.g., negative ellipticity), (2) progressive formation of lipoprotein fusion products, which are relatively depleted in apoA-I and enriched in apoA-II, apoC and lipid, and (3) apparent reversible reassociation, following exposure of HDL to high levels of GuHCl, of both apoA-II and apoC with the lipid moiety of the denatured HDL. The progressive dissociation of apoprotein from HDL produces (1) an increase in mean peak flotation rate and (2) a reduction in area of the schlieren pattern obtained upon analytic ultracentrifugation of HDL previously exposed to GuHCl. Electron microscopy indicates a progressive increase in size of the fusion products with increasing duration of exposure of the molecules to the denaturant. Perturbation of HDL structure by directly ultracentrifuging HDL in GuHCl is characterized by: (1) selective dissociation of apoA-I in the range of 2M-3M GuHCl, (2) dissociation of apoA-I, and partially of apoA-II and apoC at 4M GuHCl, and (3) extensive dissociation of apoA-I, apoA-II, and apoC at 5M GuHCl. These studies indicate differential properties of HDL apoproteins in stabilizing HDL structure and indicate a mode of lipoprotein interconversion which results from apoprotein dissociation coupled with fusion of the residual lipoprotein constituents. This interconversion process emphasizes the important role of specific apoproteins in stabilizing the lipoprotein molecule and identifies fusion as a process which can stabilize the apolar cores of lipoproteins partially depleted of their surface components.

Interconversions of HDL Induced by Dimyristoylphosphatidylcholine

Interaction of HDL with DMPC suspensions or vesicles removes some apoprotein from the lipoproteins to form discoidal complexes comparable in morphology to those described earlier in this report. The apoprotein forming the discoidal complexes appears by polyacrylamide gel electrophoresis to be primarily apoA-I. Tall and Small

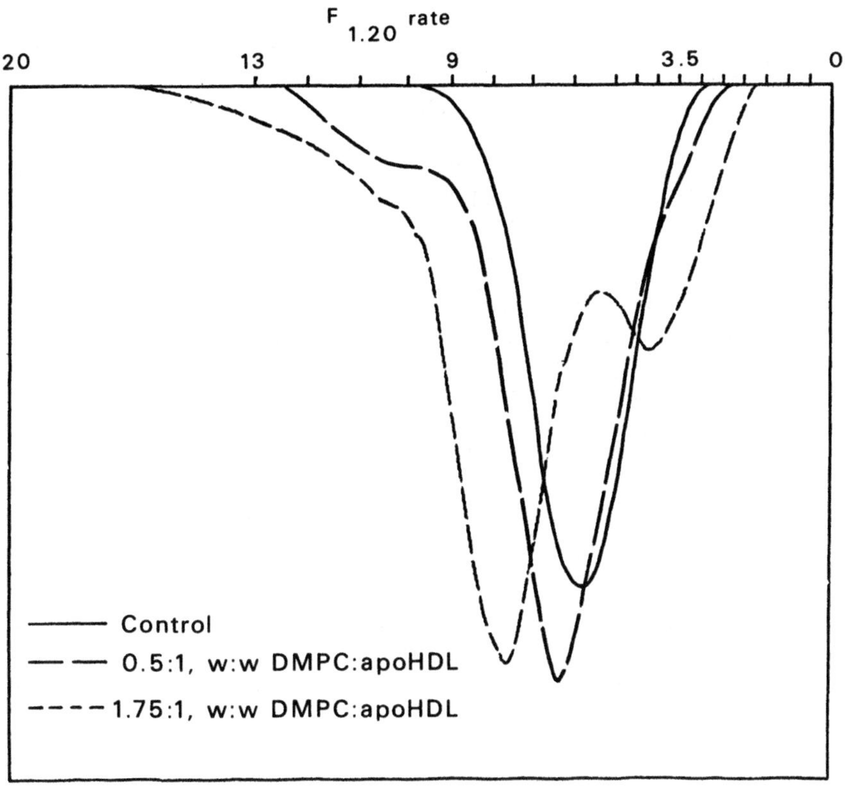

Figure 2: Analytic ultracentrifugal patterns showing the effect of interaction of HDL_{2b} with DMPC vesicles.

(27) propose that the residual lipoprotein particles, which have lost some apoA-I, fuse to form particles of increased size and decreased density. Thus, this interconversion process would be similar to that put forth to explain the changes in HDL properties after exposure of HDL to GuHCl, and comparable analytic ultracentrifugal and electron microscopic information would be expected for the fusion products resulting from interaction of HDL with DMPC. Analytic ultracentrifugation (Fig. 2) of an incubated mixture of HDL_{2b} (d 1.063-1.100) plus DMPC (1.75/1, w/w, DMPC/apoHDL; 37°C, 10 min., then cooled to transition temperature of DMPC) shows the transformation of the initial single HDL_{2b} peak into a faster main peak and a slower minor peak (28). Separate studies have established that the slow minor peak consists primarily of discoidal complexes of DMPC and apoA-I. The area of the major peak, representing the residual lipoprotein moieties after apoprotein dissociation,

does not show a decrease in apparent concentration but an increase, suggesting uptake of DMPC either by fusion products or by HDL_{2b} which has lost some apoprotein. Utilizing radioactive DMPC we have demonstrated uptake of DMPC by the residual lipoprotein moieties after interaction of HDL_{2b} with labeled DMPC vesicles. Electron microscopy of these lipoprotein species shows spherical particles which are highly deformable upon packing under the negative staining procedure used. The increased deformability is compatible with loss of surface apoprotein stabilizing the lipoprotein structure. Although the dimensions of these lipoprotein particles are somewhat larger than those of the reactant HDL_{2b}, they are significantly smaller than dimensions calculated for fusion products (e.g., dimers, etc., which have lost some surface apoprotein). An estimate of the average apoprotein loss per HDL_{2b} molecule, under the experimental conditions used, yields approximately 1 apoA-I molecule removed per HDL_{2b} molecule.

At the present time, DMPC is the only phospholipid reported to effect the above interconversion. We have observed uptake of egg yolk phosphatidylcholine by HDL species but without dissociation of surface apoprotein or discoidal complex formation. If under certain physiologic conditions long chain phospholipids might produce this interconversion in HDL, then this transformation could be of considerable implication to HDL metabolism as well as to interaction of HDL with membraneous structures.

ACKNOWLEDGEMENTS

This work was supported in part by National Institutes of Health grant HL-18574 and the U.S. Energy Research and Development Administration. The many valuable contributions of Drs. Trudy Forte and David Anderson and Ms. Elaine Gong, Pat Blanche and Dorothy Sprague are gratefully acknowledged.

REFERENCES

1. Shen, B. W., Scanu, A. M. and Kezdy, F. J. Proc. Natl. Acad. Sci. (USA) 74: 837, 1977.
2. Verdery, R. B. III and Nichols, A. V. Chem. Phys. Lipids 14: 123, 1975.
3. Nichols, A. V., Strisower, E. J., Lindgren, F. T., Adamson, G. L. and Coggiola, E. L. Clin. Chim. Acta 20: 277, 1968.
4. Eisenberg, S., Bilheimer, D. W., Levy, R. I. and Lindgren, F.T. Biochim. Biophys. Acta 326: 361, 1973.
5. Glomset, J. A., Norum, K. R., Nichols, A. V., King, W. C., Mitchell, C. D., Applegate, K. R., Gong, E. L. and Gjone, E. Scand. J. Clin. Lab. Invest. 35 Suppl. 142: 3, 1975.
6. Eisenberg, S. and Schurr, D. J. Lipid Res. 17: 578, 1976.

7. Schumaker, V. N. and Adams, G. H. J. Theor. Biol. 26: 89, 1970.
8. Olivecrona, T., Bengtsson, G., Marklund, S., Lindahl, U. and Höök, M. Fed. Proc. 36: 60, 1977.
9. Eisenberg, S. and Levy, R. I. Advan. Lipid Res. 13: 1, 1975.
10. Kane, J. P., Sata, T., Hamilton, R. L. and Havel, R. J. J. Clin. Invest. 56: 1622, 1975.
11. Mahley, R. W., Hamilton, R. L. and LeQuire, V. S. J. Lipid Res. 10: 433, 1969.
12. Hamilton, R. L. and Kayden, H. J. In: The Liver: Normal and Abnormal Functions, F. F. Becker, ed., Marcel Dekker, Inc., New York, Part A, 1974, p. 531.
13. Rubinstein, B. and Rubinstein, D. J. Lipid Res. 13: 317, 1972.
14. Forte, T., Nichols, A., Glomset, J. and Norum, K. Scand. J. Clin. Lab. Invest. 33 Suppl. 137: 121, 1974.
15. Hamilton, R. L., Williams, M. C., Havel, R. J. and Fielding, C. J. In: Atherosclerosis III, G. Schettler and A. Weizel, eds., Springer-Verlag, Berlin, abstr. 936, 1974.
16. Segrest, J. P. Chem. Phys. Lipids 18: 7, 1977.
17. Hamilton, R. L., Williams, M. C., Fielding, C. J. and Havel, R. J. J. Clin. Invest. 58: 667, 1976.
18. Fainaru, M., Havel, R. J. and Imaizumi, K. Biochim. Biophys. Acta 490: 144, 1977.
19. Mahley, R. W. and Holcombe, K. S. J. Lipid Res. 18: 314, 1977.
20. Norum, K. R., Glomset, J. A., Nichols, A. V., Forte, T., Albers, J. J., King, W. C., Mitchell, C. D., Applegate, K. R., Crony, E., Cabana, V. and Gjone, E. Scand. J. Clin. Lab. Invest. 35 Suppl. 142: 31, 1975.
21. Glomset, J. A., Norum, K. R. and King, W. J. Clin. Invest. 49: 1827, 1970.
22. Torsvik, H. Clin. Genet. 3: 188, 1972.
23. Utermann, G., Menzel, H. J. and Langer, K. H. FEBS Letters 45: 29, 1974.
24. Anderson, D. W., Nichols, A. V. and Lindgren, F. T. Biochim. Biophys. Acta, in press.
25. Anderson, D. W., Nichols, A. V., Pan, S. S. and Lindgren, F. T. Atherosclerosis, accepted for publication.
26. Nichols, A. V., Gong, E., Blanche, P., Forte, T. M. and Anderson, D. W. Biochim. Biophys. Acta 446: 226, 1976.
27. Tall, A. R. and Small, D. M. Nature 265: 163, 1977.
28. Nichols, A. V., Forte, T. M., Gong, E. L., Blanche, P. J., and Nordhausen, R., in preparation.

ASSAY OF LIPOPROTEIN LIPASE IN POSTHEPARIN PLASMA AFTER SUPPRESSION OF THE HEPATIC TRIGLYCERIDE LIPASE WITH SODIUM DODECYL SULFATE

Marietta L. Baginsky, Flora Wu, and W. Virgil Brown

Department of Medicine
School of Medicine, University of California
La Jolla, California 92093

INTRODUCTION

Lipoprotein lipase (LPL), released by heparin into the circulating blood, is thought to originate from the capillary endothelium of a variety of tissues, including muscle and adipose tissue (1). Much evidence suggests that this enzyme is of primary importance in the degradation and removal of triglyceride from plasma chylomicrons and very low density lipoproteins (2 - 4). Assay of LPL in postheparin plasma has been complicated by the presence of a second triglyceride lipase which is of hepatic origin (5 - 7). The function of this hepatic triglyceride lipase (H-TGL) in triglyceride clearance has not been elucidated, though it is possible that it plays a role in the removal of partially degradated lipoproteins following LPL action.

No substrate which is specific for LPL or H-TGL has been discovered, although marked differences in hydrolytic rates have been noted with a variety of lipids, i.e. phospholipids (8,9), monoolein (10,11) and palmitoyl-CoA (12). Therefore, determination of the activity of these enzymes has been dependent on other characteristics such as differential inhibition by chemicals, antigenic specificity, or chromatographic behavior. In the initial studies of LPL from several different tissues, Korn observed that high concentrations of NaCl (1M) or protamine sulfate markedly inhibited this activity (13). Neither of these substances inhibited the triglyceride lipase released when heparin was added to liver homogenates (5) or when liver was perfused with a heparin-containing medium (14). This observation provided the basis for an assay in which H-TGL could be determined in postheparin plasma after inhibition of LPL by protamine sulfate; LPL levels were calculated as the difference between the values obtained before and after treatment of postheparin plasma with protamine sulfate (15). Direct assay of both LPL and H-TGL has been possible following separation and partial purification of the two enzymes by heparin-Sepharose affinity chromatography (16). In this method, both enzymes are bound at low ionic strength (less than 0.4 M NaCl) to a column of Sepharose beads containing covalently linked heparin. The enzymes are then eluted with buffer containing different concentrations of sodium chloride (0.75 M for H-TGL; 1.5 M for LPL). This method permits the determination of each enzyme under its optimal assay conditions. However, it is laborious and gives falsely low values for LPL due to losses during chromatography. Selective precipitation of H-TGL or LPL with antisera prepared to the purified enzymes has provided a rapid and precise assay for each of these enzymes in human postheparin plasma (17). This method has the disadvantage of requiring generation of specific antisera for H-TGL or LPL. Also, these antisera may be species specific, thus limiting their use in comparative studies. Recent experiments in our laboratory, using purified H-TGL and LPL, have demonstrated marked differences in their stability and kinetic properties when in the presence of certain detergents (12). Sodium dodecyl sulfate (SDS) was found to inactivate H-TGL at concentrations which were protective for LPL. This observation has led to the development of a new method for the direct determination of LPL in postheparin plasma. This assay is simple and rapid, and it can be easily established using commonly available laboratory chemicals and equipment. H-TGL is also directly determined from a separate aliquot of the postheparin plasma, after inhibition of LPL by 0.75M NaCl.

METHODS AND MATERIALS

Postheparin plasma was obtained from normal and hyperlipoproteinemic volunteers after a 12-hour fast. Blood was collected (into tubes without additional anticoagulant) 15 minutes after the intravenous injection of heparin (Riker laboratories) at a dose of 60 units/Kg body weight. Plasma aliquots were kept frozen at -80°C until needed. Just prior to assay, the postheparin plasma samples were thawed and, when required, precipitated fibrin removed by a 30-min centrifugation at 270 x g. For H-TGL determination, the plasma sample was diluted with an equal volume of 0.2 M Tris-HCl buffer, pH 8.2, and assayed immediately afterwards. Unless otherwise indicated, LPL was determined after pre-incubation (26°C) of the postheparin plasma diluted with an equal volume of 100mM SDS in 0.2 M Tris-HCl, pH 8.2, for the time periods indicated in the legend of figures. Pre-incubation conditions in an experiment using purified enzymes are given in the legend of Fig. 1. Assays were done in triplicate in a final volume of 0.5 ml. Hydrolysis of glyceryl trioleate by H-TGL was measured at pH 8.8 in 0.2 M Tris-HCl - 0.75 M NaCl, no activator serum added. LPL activity was measured at pH 8.2 in 0.2 M Tris-

HCl - 0.10 M NaCl buffer, plus 90 μl of serum as source of activator apolipoprotein(s). The assay medium for both enzymes contained per ml: 9 μmoles glyceryl tri ($1-^{14}C$)-oleate (SA = 0.088 μC/μmole) (emulsified by sonification in 5% gum arabic in 0.2 M Tris of the appropriate pH), 100mg of fatty acid-free bovine serum albumin (18) and 15mg of gum arabic. The assay was started by the addition of 10 μl of the postheparin plasma solutions described above. After 60 minutes of incubation at 28°C the reaction was terminated and the released fatty acid determined as described by Huttenen, et al. (19), with the following exception: elution of the fatty acid from the resin was done by incubation for 90 minutes at 70°C with 1ml of an equal mixture of 2 M KOH in methanol and Triton X-100. Results were corrected for recovery as determined using standard solutions of $1-^{14}C$-oleic acid and were expressed in μmoles free fatty acid liberated in an hour, per ml of postheparin plasma. Unlabeled glyceryl trioleate was from Nu-Check Prep, Inc., glyceryl tri ($1-^{14}C$)-oleate from Dhom, SDS from Eastman (lot # 175X), Trizma Base and gum arabic were from Sigma Chemical Company.

RESULTS

The effect of SDS on purified H-TGL and LPL. - Previous studies had indicated that purified H-TGL was inactivated (99%) when pre-incubated with 1 mM SDS for 5 minutes at 26°C. LPL activity, under the same conditions, was almost completely preserved (12). The possible use of SDS for the selective inhibition of H-TGL in postheparin plasma was investigated in a model system prepared by the addition of pre-heparin plasma to each of the purified enzymes (Fig. 1). In this experiment, complete inhibition of H-TGL was also obtained in the presence of added plasma protein, although higher concentrations of SDS were required than with purified enzymes alone. The higher amount of SDS needed to obtain the same effect is consistent with the known interaction of protein and detergent on a weight to weight basis (20).

As reported previously, LPL is inactivated when diluted in a buffer without detergent (12). This was also found to be the case here even in the presence of serum. However, LPL activity was fully protected (and somewhat activated) when incubated in the presence of SDS. LPL remained fully active at an SDS concentration of 25mM in the pre-incubation mixture (yielding 2mM in the assay system). This concentration of SDS almost completely inactivated H-TGL.

Pre-incubation of postheparin plasma with SDS. - Using postheparin plasma from normal donors, the lipase activity assayed under conditions optimal for H-TGL declined progressively with pre-incubation at increasing concentrations of SDS (Fig. 2). Virtually no activity remained at 50mM. When these plasma samples were assayed under conditions optimal for LPL (0.1M NaCl, plus added serum) a decline in activity was also observed. However,

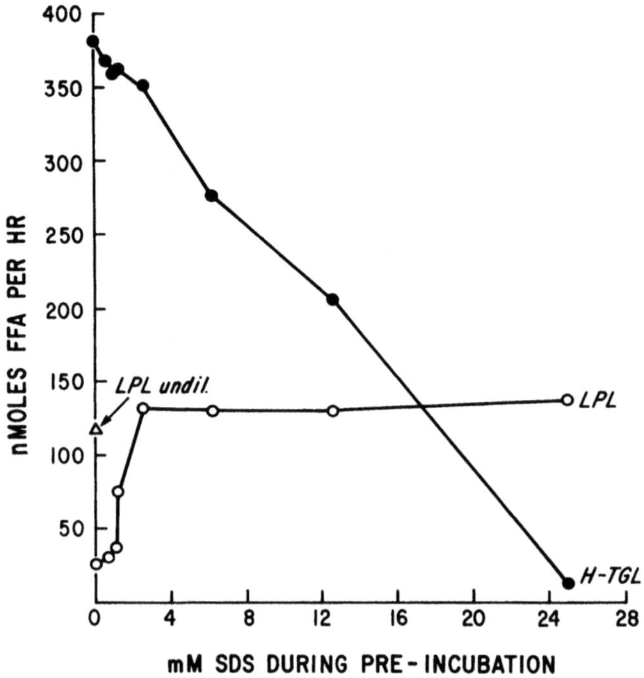

Figure 1: Effect of SDS on purified triglyceride lipases. H-TGL and LPL were isolated from the postheparin plasma of a normal volunteer by heparin-Sepharose chromatography (25). Both enzymes in this experiment were used as obtained from the column (i.e., H-TGL in 0.005 M veronal buffer containing 0.75 M NaCl and LPL in veronal buffer with 1.5 M NaCl, both of pH 7.0). Pre-incubation mixtures for H-TGL contained: H-TGL solution (480 µg protein per ml), veronal buffer - 1.5 M NaCl, normal serum and SDS (at double the concentration indicated in the abscisa, in 0.2 M Tris-HCl of pH 8.2) in the ratios (v/v) of 1:2:1:4. The pre-incubation mixtures for LPL contained: LPL solution (132µg protein/ml), veronal buffer -0.75 M NaCl, normal serum and SDS solutions (as above) in the ratios of 2:1:1:4. Samples were pre-incubated for 10 minutes at 26°C, after which activity was determined as indicated in the Methods section, using 40 µl of enzyme solution. Values given in the ordinate are expressed in nmoles FFA liberated per hour per ml of undiluted enzyme solution.

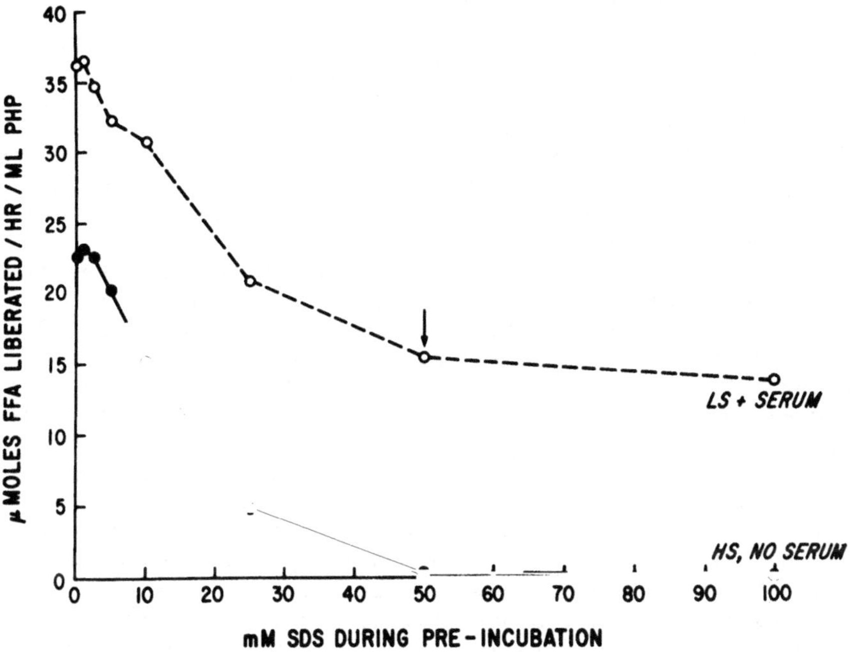

Figure 2: Effect of SDS concentration on H-TGL and LPL activities of postheparin plasma. Postheparin plasma (500 µl) from a normal human donor was mixed with an equal volume of SDS solution in 0.2 M Tris, pH 8.2,at concentrations to provide those indicated on the abscisa. After 10 minutes at 26°C, assays for glyceryl trioleate hydrolase activities were done as described in the Methods section. The values indicated by LS + serum (O---O) were obtained using the assay medium optimal for LPL; the curve indicated by HS, no serum (●—●) was obtained with assays done under the conditions optimal for H-TGL. The arrow indicates the SDS concentration (50mM) chosen for all subsequent studies reported in this paper.

a plateau was reached which, in the example shown in Fig. 2, was at approximately 1/2 the original value. This loss in activity is consistent with inactivation of H-TGL, an enzyme that is also determined (although sub-optimally) under the low salt condition. The activity measured at concentrations of SDS of 50mM or above, is thus believed to be due entirely to LPL.

Effect of duration of pre-incubation with SDS on enzyme activities. Postheparin plasma samples from a normal subject and from one patient each with Type I and Type V hyperlipoproteinemia were compared during pre-incubation with 50mM SDS (Fig. 3). The H-TGL activity (●—— ●) fell rapidly with a decline evident even after 10 seconds of pre-incubation and, in the normal, no residual activity was detectable after approximately 20 minutes (Fig. 3A). When H-TGL activity had been suppressed to negligible levels, activity assayed under low salt conditions (0 --- 0) reached a constant value.

Postheparin plasma samples of Type I subjects assayed by other methods have been found to be markedly deficient in LPL. H-TGL is present, however, in subjects with this disorder (21-23). This was confirmed in the postheparin plasma sample shown in Fig. 3B, using both heparin-Sepharose chromatography and immunochemical techniques (data not given). The complete loss of glyceryl trioleate hydrolase activity by pre-incubation with SDS of this plasma, known to be lacking in LPL, is again consistent with inactivation of H-TGL by SDS.

The experiment depicted in Fig. 3C demonstrates the effect of pre-incubation with SDS of postheparin plasma from a subject with Type V hyperlipoproteinemia. Activity measureable under the assay conditions for H-TGL was still evident at 30 minutes in the presence of 50mM SDS. This may be due, in part, to the very high initial H-TGL activity found in this patient's plasma. This slower rate of inactivation was also found in other hyperlipoproteinemic plasma samples with lower H-TGL levels. Thus, protection of H-TGL by the lipoproteins or lipid-detergent interactions could also be a cause for this slower inactivation. Extending the pre-incubation period to assure complete loss of H-TGL activity raised, however, the possibility of producing serious deterioration in LPL activity. Shown in Fig. 4 are results from studies using postheparin plasma samples from subjects with Type IV hyperlipoproteinemia and a normal individual, in which the pre-incubation period was extended to 45 minutes. Whereas the H-TGL activity approached negligible values, the activity measured in low ionic strength and with serum activator was either stable (Fig. 4A) or increased (Figs. 4B and 4C) with the more prolonged pre-incubation in SDS.

The studies using postheparin plasma from two subjects with Type V hyperlipoproteinemia is shown in Fig. 5. In these experiments, the pre-incubation with 50mM SDS was extended to one hour. Note that the LPL activity did appear to reach a plateau at the later time points. However, the prolonged survival of activity

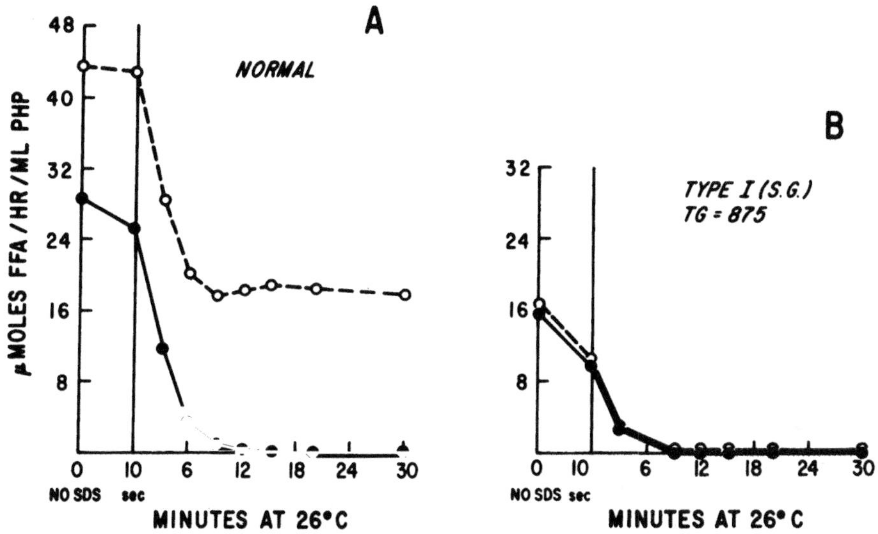

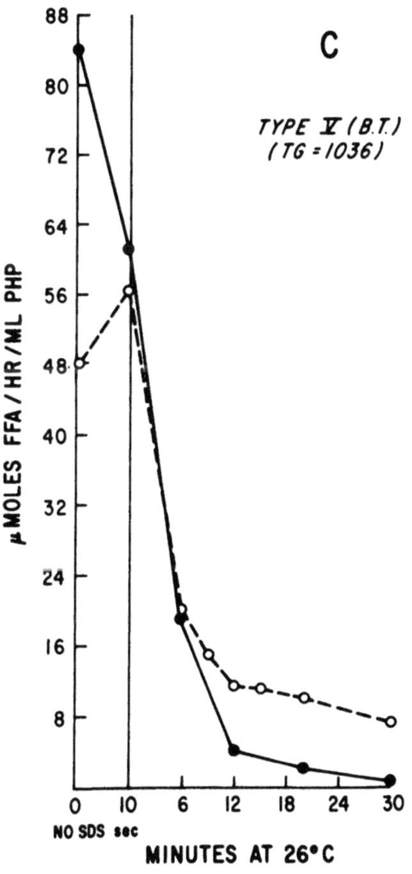

Figure 3: Glyceryl trioleate hydrolase activities remaining after pre-incubation with SDS. Postheparin plasma was pre-incubated with SDS (50mM final concentration), as described in the text. At the times indicated on the abscisa, aliquots were assayed under the conditions optimal for LPL (0---0) and for H-TGL (● — ●). The plasma was obtained from a normal volunteer (Figure 3A) and subjects with Type I (Figure 3B) and Type V hyperlipoproteinemia (Figure 3C).

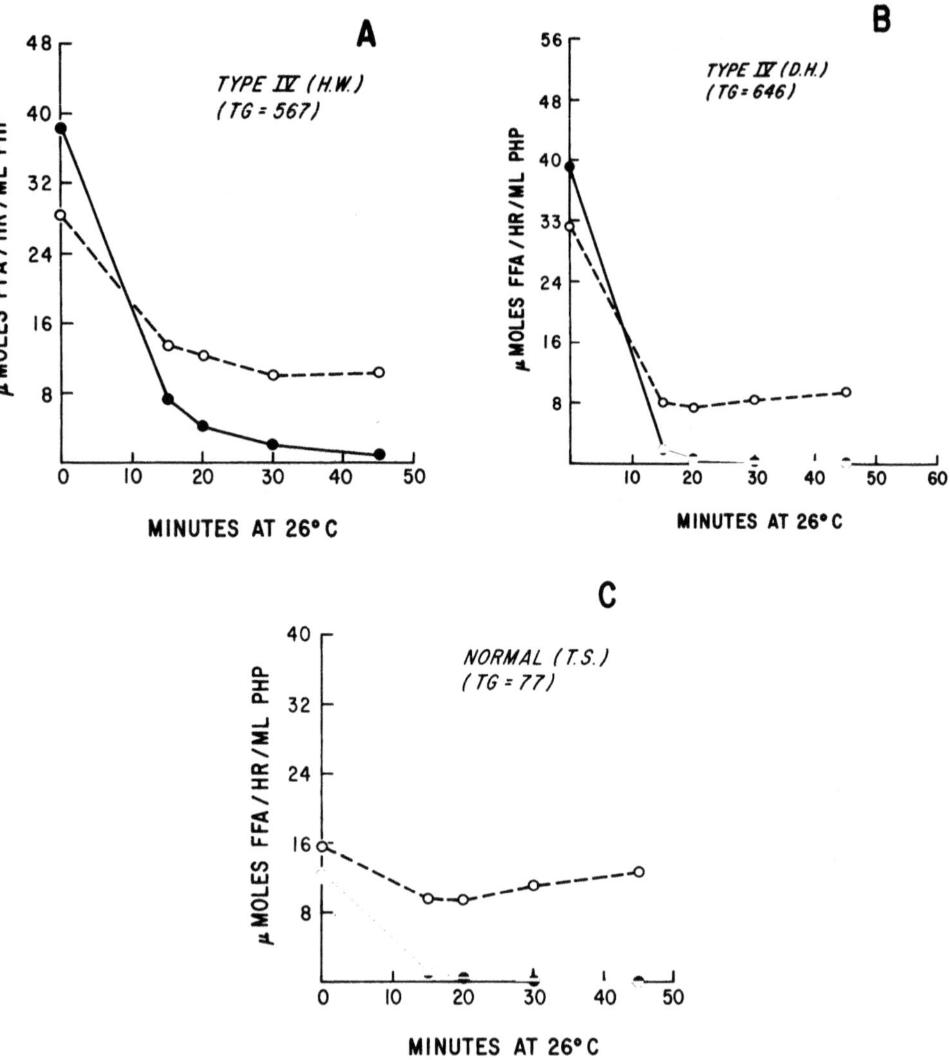

Figure 4: Glyceryl trioleate hydrolase activities remaining after 45 minutes of pre-incubation. Pre-incubation of postheparin plasma was carried out as described in the legend to Figure 3. At the times indicated on the abscisa assays were conducted under conditions optimal for LPL (0---0) and for H-TGL (●—●), as described in the Methods section. The values obtained with postheparin plasma from two subjects with Type IV hyperlipoproteinemia are shown in Figures 4A and 4B, and those from a normal subject in Figure 4C.

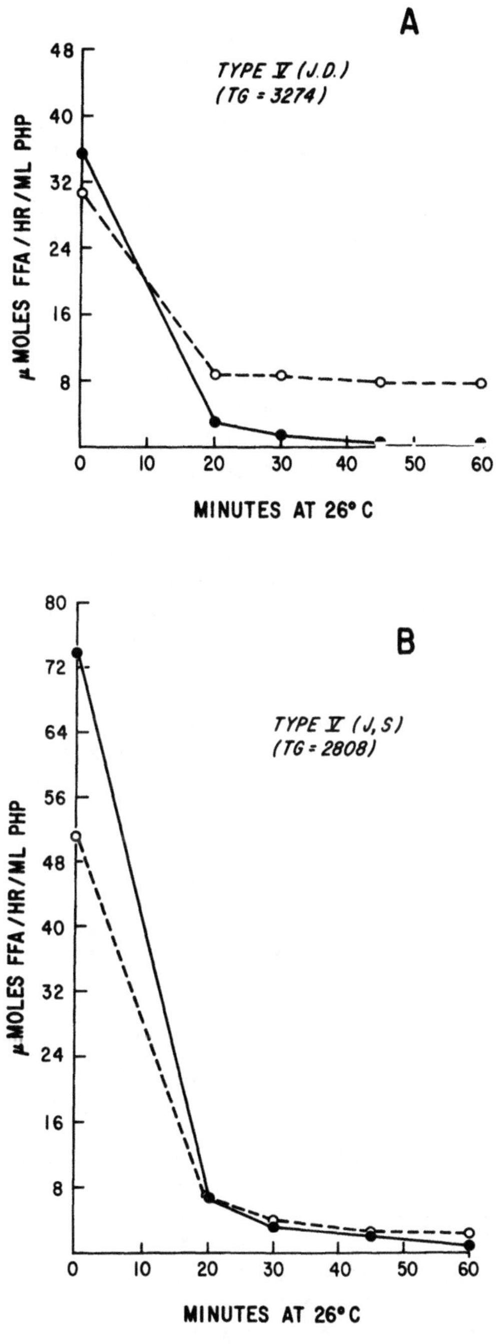

Figure 5: Effect of pre-incubation with SDS on glyceryl trioleate hydrolase activities of postheparin plasma from subjects with Type V hyperlipoproteinemia. The conditions of pre-incubation and assay were conducted as described in the legend of Figure 3. Subject A.D. has primary familial Type V hyperlipoproteinemia and subject J.S. has Type V hyperlipoproteinemia and insulin dependent diabetes mellitus. 0---0, assays done under conditions optimal for LPL; ●—●, assays done under conditions optimal for H-TGL.

in high salt concentrations continued to be a problem in certain patients as seen in Fig. 5B. Recently, pre-incubation experiments have been conducted for longer periods. The activity measured using assay conditions for LPL was maintained for at least 2 hours in these samples, while the H-TGL was completely inactivated. In the few postheparin plasma samples from Type V subjects examined thus far, it is clear that some have LPL activity within normal limits (Fig. 5A) while others have low activity (Fig. 5B). Of considerable interest is the very high H-TGL activity estimated in several of these plasma samples from subjects with Type V hyperlipoproteinemia.

DISCUSSION

Differing effects of low concentrations of detergents on the tertiary structure of a variety of proteins has been described (24). It is surprising, however, that two enzymes which appear to differ only in their carbohydrate side chains (25) could be so clearly distinguished in their response to exposure to SDS solutions. The stabilization of LPL may result, in part, from prevention of aggregation. The appearance of turbidity in solutions of purified LPL on reduction of the ionic strength by dilution has been repeatedly observed in our laboratory, and is consistent with a very high tendency for aggregation. When SDS is contained in the diluting buffer solutions, this turbidity is prevented and activity of the enzyme is fully maintained.

The selective inactivation of H-TGL is seen with both the purified enzyme and in plasma, over a broad range of lipoprotein values. With lipemic plasma samples, higher concentrations of SDS or more prolonged exposure is required to produce the same percentage of inactivation than with postheparin plasma from normal individuals.

The data presented in this article represents a preliminary report of the basis for a promising method for the direct assay of LPL in postheparin plasma. Conditions have been established empirically to completely inhibit all triglyceride hydrolase activity with the known assay characteristics of H-TGL. Using assay conditions defined for LPL, glyceryl trioleate hydrolase activity declined with the H-TGL inactivation, reaching a plateau which was stable for at least 2 hours of pre-incubation, under the conditions used. The remaining activity is believed to be due to LPL for the following reasons: 1) The purified LPL showed similar stability in the presence of pre-heparin plasma (Fig. 1) and no LPL activity was observed when this assay was done at high ionic strength (data not shown); 2) Similarly, the SDS-resistant activity of postheparin plasma was inhibited by 0.75 M NaCl (Fig. 2); and 3) Postheparin plasma from a subject with Type I hyperlipoproteinemia, (known to lack LPL) showed no glyceryl trioleate hydrolase activity after pre-incubation with SDS (Fig. 3B).

Initial results from a study currently underway suggest a

very close correlation between LPL, as assayed following immunoprecipitation of H-TGL, and as assayed after complete inhibition of the latter enzyme by pre-incubation with SDS. The values in the latter method are 10-15% higher, perhaps reflecting some stabilization or activation of LPL in the presence of detergent. Similar comparative studies with the LPL recovered from the heparin-Sepharose affinity columns also showed a good correlation. However, the latter values were 40-50% lower than those obtained with the SDS method. The difference may reflect the summation of LPL activation by SDS and the fractional loss of LPL during chromatography. On the other hand, values for H-TGL, were almost identical with the three methods.

Using the same quantity of SDS, different degrees of inhibition of H-TGL were obtained with different batches of this detergent. Therefore, the concentration of SDS needed for complete inhibition of H-TGL in postheparin plasma, as well as time of pre-incubation required, have not yet been definitely set. This is particularly true with plasma samples containing high lipoprotein levels.

The acceptance of the values for LPL and H-TGL by the method proposed in this communication is dependent on a clear demonstration of the precision of the procedure and accuracy of the values obtained among the various forms of hyperlipoproteinemia. The definition of accuracy must be made in a comparative sense using the best method of assay now available. Such studies are currently underway in our laboratory using the selective immunochemical precipitation of the two enzymes as the standard procedure for comparison.

ACKNOWLEDGEMENTS

The excellent assistance of Mrs. Florence Casanada is greatly appreciated.

This work was supported by the Specialized Center of Research on Atherosclerosis (SCOR) as funded by the National Heart, Lung and Blood Institute through grant #HL 14197-06.

REFERENCES

1. Robinson, D.S. and J.E. French. 1960. Heparin, the clearing factor lipase, and fat transport. Pharmacol. Rev. 12: 241-263.

2. Scow, R.O., J. Blanchette-Mackie and L.C. Smith. 1976. Role of capillary endothelium in the clearance of chylomicrons. A model for lipid transport from blood by lateral diffusion in cell membranes. Circulation Res. 39: 149-162.

3. Kopiang, I.P., A. Bensadoun, and M.W. Yang. 1976. Effect of an anti-lipoprotein lipase serum on plasma triglyceride removal. J. Lipid Res. 17: 498-505.

4. Jackson, R.L., J.D. Morrisett, and A.M. Gotto, Jr. 1976. Lipoprotein structure and metabolism. Physiological Rev. 56: 259-299.

5. La Rosa, J.C., R.I. Levy, H.G. Windmueller, and D.S. Fredrickson. 1972. Comparison of the triglyceride lipase of liver, adipose tissue, and post-heparin plasma. J. Lipid Res. 13: 356-363.

6. Greten, H., A.D. Sniderman, J.G. Chandler, D. Steinberg, and W.V. Brown. 1974. Evidence for the hepatic origin of canine postheparin plasma triglyceride lipase. FEBS Letters 42: 157-160.

7. Ehnholm, C., W. Shaw, H. Greten, W. Lengfelder and W.V. Brown. 1974. Separation and characterization of two triglyceride lipase activities from human postheparin plasma. In: Atherosclerosis III. Proceedings 3rd. International Symposium. G. Schetler and A. Winzer, Eds. Springer-Verlag, Berlin -Heidelberg -New York, NY, 557-560.

8. Ehnholm, C., W. Shaw, H. Greten and W.V. Brown. 1975. Purification from human plasma of a heparin-released lipase with activity against triglyceride and phospholipids. J. Biol. Chem. 250: 6756-6761.

9. Brown, W.V., W. Shaw, M. Baginsky, J. Boberg and J. Augustin. 1976. In: Lipoprotein Metabolism, H. Greten, Ed. Springer-Verlag, Berlin - Heidelberg - New York, NY, 2-6.

10. Greten, H., B. Walter and W.V. Brown. 1972. Purification of a human post-heparin plasma triglyceride lipase. FEBS Letters 27: 306-310.

11. Nilsson-Ehle, P., T. Egelrud, P. Belfrage, T. Olivecrona and B. Borgström. 1973. Positional specificity of purified milk lipoprotein lipase. J. Biol. Chem. 248:6734-6737.

12. Baginsky, M.L. and W.V. Brown. 1977. Differential characteristics of purified hepatic triglyceride lipase and lipoprotein lipase from human postheparin plasma. J. Lipid Res. 18: XX.

13. Korn, E.D. 1955. Clearing factor, a heparin-activated lipoprotein lipase. I. Isolation and characterization of the enzyme from normal rat heart. J. Biol. Chem. 215: 1-14.

14. Hamilton, R.L. 1965. Post-heparin plasma lipase from the hepatic circulation. Diss. Abstr. 26: 24.

15. Krauss, R.M., H.G. Windmueller, R.I. Levy and D.S. Fredrickson. 1973. Selective measurement of two different triglyceride lipase activities in rat postheparin plasma. J. Lipid Res. 14: 286-295.

16. Boberg, J., J. Augustin, M.L. Baginsky, P. Tejada, and W.V. Brown. 1974. Quantitative determination of hepatic and lipoprotein lipase activities from human post-heparin plasma. Circulation 50: Suppl. III, 21.

17. Huttunen, J.K., C. Ehnholm, P.K. Kinnunen and E.A. Nikkilä. 1975. An immunochemical method for the selective measurement of two triglyceride lipases in human postheparin plasma. Clin. Chim. Acta 63: 335-347.

18. Chen, R.F. 1967. Removal of fatty acids from serum albumin by charcoal treatment. J. Biol. Chem 242: 173-181.

19. Huttenen, H.K., J. Ellingboe, R.C. Pittman, and D. Steinberg 1970. Partial purification and characterization of hormone-sensitive lipase from rat adipose tissue. Biochim. Biophys. Acta 218: 333-346.

20. Reynolds, J.A. and C. Tanford. 1970. The gross conformation of protein-sodium dodecyl sulfate complexes. J. Biol. Chem. 245: 5161-5165.

21. Krauss, R.M., R.I. Levy, and D.S. Fredrickson. 1974. Selective measurement of two lipase activities in postheparin plasma from normal subjects and patients with hyperlipoproteinemia. J. Clin. Invest. 54: 1107-1124.

22. Greten, H., R. DeGrella, G. Klose, W. Rascher, J.L. deGennes, and E. Gjone. 1976. Measurement of two plasma triglyceride lipases by an immunochemical method: Studies in patients with hypertriglyceridemia. J. Lipid Res. 17: 203-210.

23. Brown, W.V., M.L. Baginsky, and C. Ehnholm. (In press). Primary Type I and Type V hyperlipoproteinemia. In: Hyperlipidemia: Diagnosis and Therapy. Rifkind, B. and R.I. Levy; (Eds.) Grune and Stratton, New York, NY.

24. Steinhart, J. and J.A. Reynolds. 1969. Multiple equilibria in proteins. Academic Press, New York, NY; 234-302.

25. Augustin, J., H. Freeze, J. Boberg, and W.V. Brown. 1976. Human postheparin plasma lipolytic activities. In: Lipoprotein Metabolism. H. Greten (Ed). Springer-Verlag, Berlin - Heidelberg - New York, NY: 7-12.

PHOSPHOLIPASE A_2 INTERACTION WITH VARIOUS PHOSPHOLIPIDS.

A MODEL FOR THE LIPOPROTEIN MOLECULE?

G.H. de Haas, A.J. Slotboom, H.M. Verheij, J.C. Vidal,
E.H.J.M. Jansen and H. Baartmans
Laboratory of Biochemistry, State University of Utrecht
Transitorium 3, University Centre "De Uithof"
Padualaan 8, Utrecht, The Netherlands

INTRODUCTION

This NATO-workshop is devoted to the structure and function of the lipoprotein molecule. Important building stones of these complexes are the so-called "apoproteins" and phospholipids. Considerable progress has been made in the isolation and elucidation of the structure of proteins and phospholipids. One of the major problems is how the lipids and proteins interact with each other on a molecular level. Usually reconstitution experiments with pure apoproteins and different types of phospholipids are monitored by calorimetric and spectroscopic techniques such as fluorescence, ultraviolet difference spectroscopy or NMR. However, no sensitive and easy bioassay is available to control the structure of the reconstituted lipoprotein. In addition, the tendency of several apoproteins to aggregate in solution, complicates the interaction study. Therefore comparison might be useful with a modelsystem in which a small water-soluble protein molecule interacts with lipid aggregates, especially if the interaction can be followed by a change in enzyme activity. In this respect lipolytic enzymes are attractive: they are usually small, rather stable proteins, which can easily be obtained in a pure state, and, as we will see, the specific interaction of these proteins with certain lipid-water aggregates is often accompanied by an exponential increase in enzyme activity.

The purpose of the present study is to investigate the binding forces between a lipolytic enzyme and its water-insoluble substrate and to try to answer such questions as:

- which part of the protein molecule is involved in the interaction?
- should the interaction be defined as Adsorption, Anchoring or Penetration?

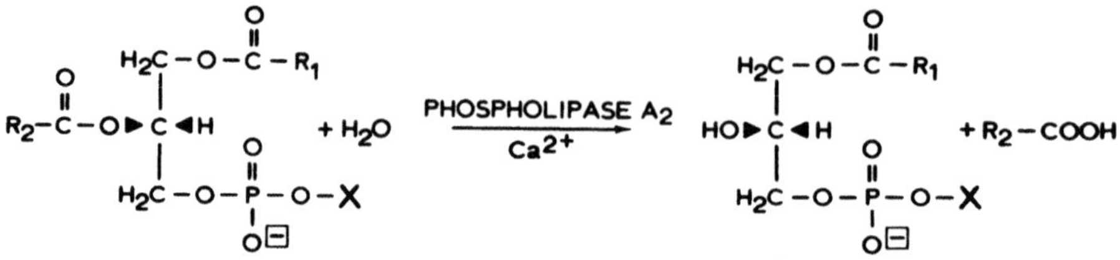

Fig. 1. *Reaction catalyzed by phospholipase A_2. X represents any of the naturally occurring moieties found in 3-sn-phosphoglycerides (e.g., choline, ethanolamine, glycerol, etc.). R_1 and R_2 are long alkyl chains.*

Fig. 2. *Revised primary structure of porcine pancreatic (pro)phospholipase A_2. (From Puijk et al. 10.)*

-is the interaction reversible?
-which parameters of the lipid-water interface control the interaction?

Most of our work has been done with mammalian pancreatic phospholipases A_2 (EC 3.1.1.4). This enzyme has been isolated from ox, pig, horse and sheep pancreas and catalyzes the hydrolysis of the 2-acyl ester linkage in naturally occurring phosphoglycerides (1). (See Figure 1). The enzyme acts highly stereospecific and only 3-*sn*-phosphoglycerides are degraded whereas the stereoisomeric 1-*sn*-phosphoglycerides (D-α-phosphatides) are competitive inhibitors characterized by the same binding constant to the enzyme as the 3-*sn*-analogs (2). The pancreatic enzymes have an absolute requirement for Ca^{2+} ions which bind in a 1:1 molar ratio to the enzyme at the catalytic or high-affinity site. Ba^{2+} and Sr^{2+} ions bind to this site with a similar affinity constant as Ca^{2+}, but no hydrolysis takes place. Mg^{2+} ions do not bind to the enzyme (3). Sofar only a few transition metals such as Gd^{3+} and Tb^{3+} were found to replace Ca^{2+} with retention of some enzyme activity (4).

The mammalian pancreatic phospholipases A_2 are all secreted by the pancreas in a zymogen form, probably to avoid uncontrolled degradation of membraneous material at the site of biosynthesis (5-9). The amino acid sequences of the porcine (10), bovine (11) and equine (12) phospholipase A_2 have been determined. Figure 2 shows the primary structure of the porcine zymogen, a protein whose X-ray structure has been determined recently by Drenth and colleagues (13). (Compare figures 10^a and 10^b).

The zymogen consists of a single polypeptide chain of 131 amino acids cross-linked by 7 disulfide bridges. During conversion into the active enzyme by limited proteolysis with trypsin the N-terminal $Arg-Ala^1$ bond is split and the activation heptapeptide PyroGlu-Glu-Gly-Ile-Ser-Ser-Arg is liberated. After activation there is a conformational change which can be monitored by fluorescence spectroscopy of the unique Trp^3 residue located at the N-terminal site (14). The newly generated $\alpha\text{-}NH_3^+$ group of the N-terminal $L\text{-}Ala^1$, which has a pK of 8.3, most probably forms an internal ion-pair with a buried carboxylate function in the interior of the protein (15).

The natural substrates of phospholipase A_2 are long-chain phospholipids which are insoluble in water and aggregate forming double-layered liposomes or vesicles. Taking into account that usually the lipolytic enzymes are small molecules and water-soluble, it is clear that lipolysis is a form of heterogeneous catalysis (17). The hatched area in figure 3 represents the lipid-water interface of the insoluble lipids. (= surface of the micelle, liposome, vesicle or monolayer at the air-water interface). The bulk aqueous phase contains the highly soluble enzyme E. Therefore lipolysis consists at least of two steps:

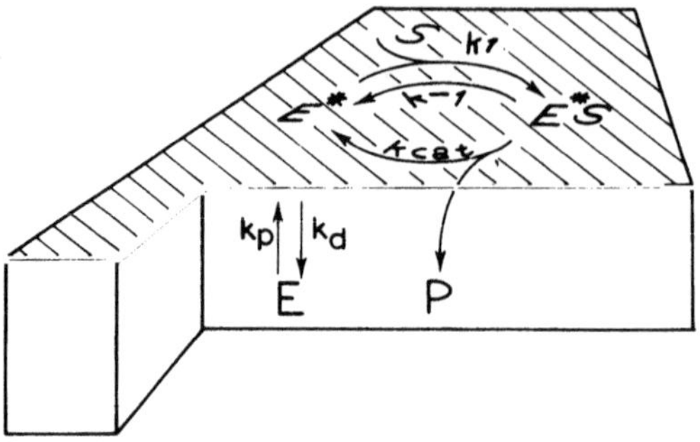

Fig. 3. *Model for the action of a soluble enzyme at an interface. (From Verger et al. 16.)*

-the enzyme has to interact first with the lipid-water interface: E ⇄ E*. We assume for the moment being that this interaction is reversible and characterized by the rate constants k_p and k_d. Dependent on the extent of interaction this first step can be called Adsorption, Anchoring or Penetration. Most probably the enzyme will change its conformation in the new environment, denoted as E*.

-Once the enzyme is located in the interface, we assume a "normal", two-dimensional Michaelis-Menten equilibrium: one molecule of protein, being "dissolved in a sea of rapidly moving monomer lipid molecules" (18-20), will react in a reversible way with a lipid molecule S forming the activated complex E*S. The E*S complex decomposes into product P and enzyme E*. This model implies that the activity of a lipolytic enzyme acting on a lipid aggregate is regulated not only by the architecture of its catalytic site, but will be strongly dependent also on the first step E ⇄ E*. Therefore it may be expected that a study of the kinetics of the enzyme will provide valuable information on the interaction process E ⇄ E*.

As will be seen later on, it is difficult to use in kinetic studies natural long-chain phospholipids as substrate. However, upon acyl-chain shortening below nine carbon atoms, lecithins form micelles in stead of multilayered liposomes and enzymic degradation of these micellar phospholipids can easily be quantitated by a number of techniques. As shown in Figure 4, using as substrate 1,2-diheptanoyl-*sn*-glycero-3-phosphorylcholine, both the active phospholipase A_2 (curve a) and the zymogen (curve b) follow normal Michaelis-Menten kinetics up to the critical micellar concentration (CMC) of the lecithin. This indicates that both proteins are able to hydrolyze monomeric substrate at comparable, but slow rates. Therefore, it has to be concluded that in both proteins a similar active site is present! (21). Kinetic difference becomes apparent upon substrate aggregation: when micelles are formed, a tremendous increase in en-

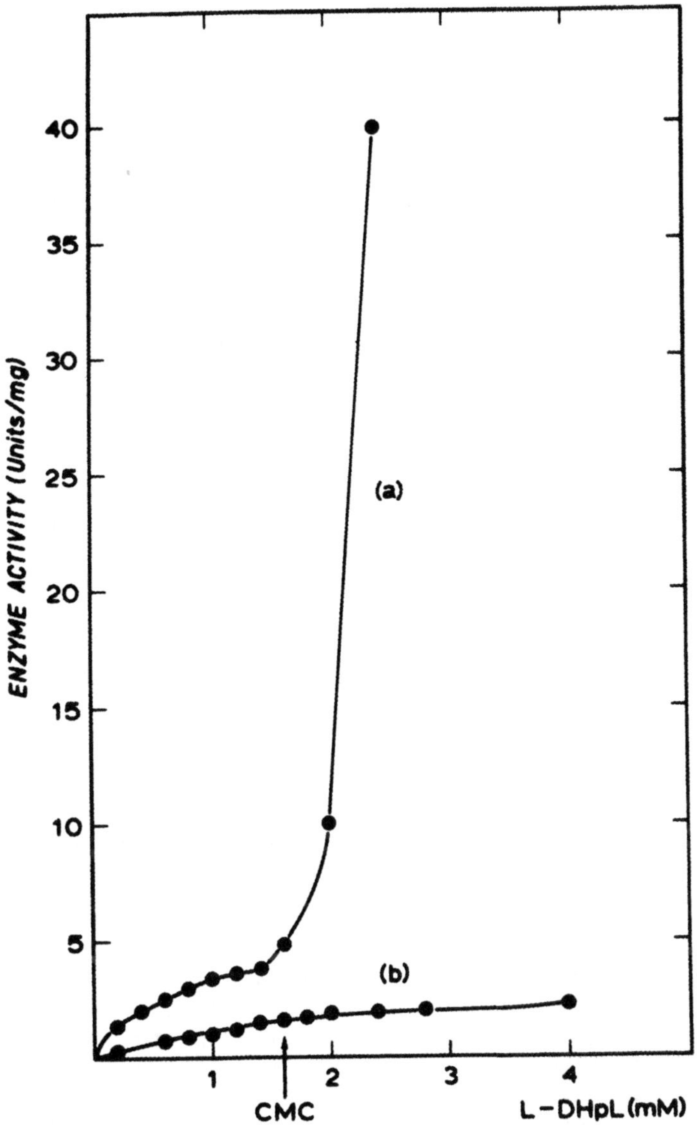

Fig. 4. *"Michaelis" curves showing the activity of phospholipase A_2 (curve a) and its zymogen (curve b) as a function of 1,2-diheptanoyl-sn-glycero-3-phosphorylcholine (= L-DHpL) concentration. Assay conditions: 0.5 mM NaAc, 0.1 M NaCl, pH 6.0, 40°. CMC stands for critical micellar concentration. (From Pieterson et al. 21.)*

zymic activity is observed for the active enzyme but not for the zymogen. Direct binding studies such as equilibrium gel filtration (21), ultraviolet difference spectroscopy (22), fluorescence measurements (22) and microcalorimetry (23) have shown that, in contrast to the active enzyme, the zymogen is not able to bind to lipid-water interfaces. The difference between both proteins has been attributed to the presence of a particular surface domain in the active enzyme through which it recognizes and binds to certain lipid-water interfaces (Fig. 3 E $\rightleftarrows$ E*). This site, the so-called Interface Recognition Site (IRS) is not present in the zymogen. From the above it can be concluded that there is a functional difference between the IRS and the active site area where *monomeric* substrates and Ca^{2+} are bound and hydrolysis occurs (*c.f.* Fig. 3).

Relative Location of the Active Site and IRS

As has been shown previously (24), porcine pancreatic phospholipase A_2 is irreversibly inactivated by *p*-bromophenacylbromide. This reagent specifically reacts with His^{48}. The fact that Ca^{2+} ions and monomeric substrate analogs give a nearly absolute protection against this modification seems to indicate that the reagent can be considered as an active site-directed inhibitor. Direct proof that the chemical modification is preceded by a reversible complex formation between enzyme and inhibitor could not be obtained for the porcine enzyme, probably because of a rather weak affinity of this protein for *p*-bromophenacylbromide. Inactivation experiments of the bovine enzyme as a function of the inhibitor* concentration, however, clearly show a saturating character (Figure 5), and a dissociation constant K_i = 3 mM was found at pH 5.5 (25). Also in this case it could be demonstrated that only His^{48} was modified and again protection was observed by Ca^{2+} ions and monomeric substrates. So it can be concluded that His^{48} is one of the active site residues of phospholipase A_2 and that binding of the essential ligands Ca^{2+} and monomeric lecithin takes place close to this residue. Although the *p*-bromophenacylbromide-inhibited phospholipase A_2 does not possess any enzyme activity, indicating a complete loss of active center, it has been demonstrated (2) that the modified protein still interacts with lipid-water interfaces. Mixtures of native and His^{48}-modified phospholipase A_2 have been studied in competition experiments for a micellar lipid-water interface by equilibrium gel filtration.

---

* Because of a higher solubility phenacylbromide was used in stead of *p*-bromophenacylbromide.

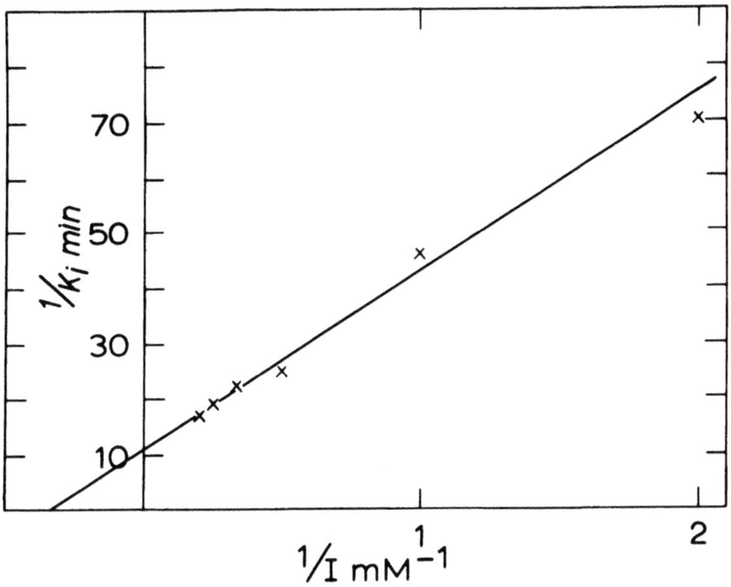

Fig. 5. *Double reciprocal plot of the inactivation velocity as a function of the phenacylbromide concentration. Conditions: 0.1 M sodium cacodylate-HCl (pH 5.5) containing 0.1 M NaCl; bovine phospholipase concentration 2 x 10^{-4} M. T = 25°C. (From Dutihl, 25).*

Treatment of the data according to Best-Belpomme and Dessen (26) yields the following equation:

$$\left[\frac{1}{LE} - \frac{1}{LE_0}\right] = \frac{K_M}{K_E} \cdot \frac{[M]}{[E]}$$

in which K_E and K_M are the affinity constants of the native and modified enzyme, respectively. [M] and [E] are the concentrations of unliganded modified and native phospholipase A2, respectively. LE is the fraction of sites combined with native phospholipase A_2, LE_0 is the fraction of sites combined with native phospholipase A2, in the absence of modified phospholipase A2.

Therefore a plot of $\left[\frac{1}{LE} - \frac{1}{LE_0}\right]$ as a function of $\frac{[M]}{[E]}$ should yield

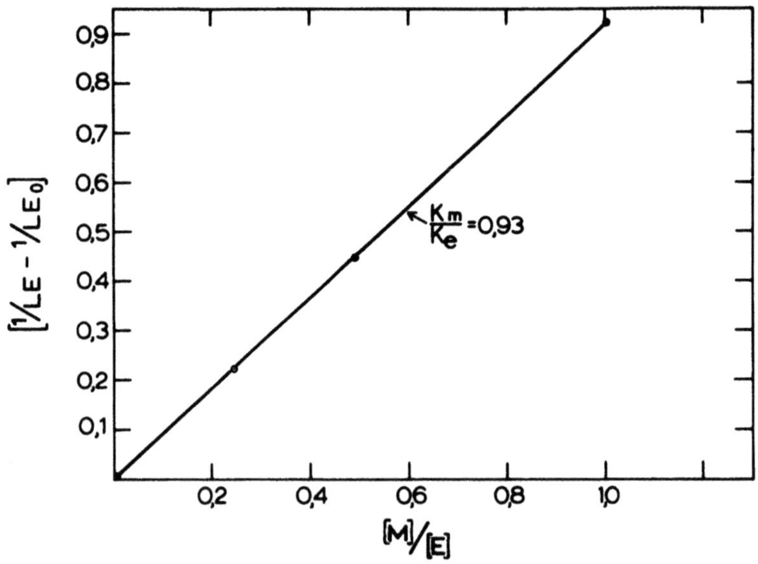

Fig. 6. *Plot of* $[1/LE - 1/LE_0]$ *vs* $[M]/[E]$ *according to the method of Best-Belpomme and Dessen (26) showing identical affinities of phospholipase* A_2 *and p-bromophenacylbromide inhibited phospholipase* A_2 *for micellair phospholipids (21).*

a straight line through the origin, the slope of which represents the ratio of the affinity constants. The result is given in Figure 6. It can be concluded that blocking of the active site in phospholipase A_2 has no influence on the properties of the IRS. This implies that the IRS is not only functionally but also *topographically* distinct from the active site.

Which Amino Acids are involved in the IRS?

The intriguing question where the IRS is located on the enzyme surface has been investigated by a variety of techniques: Direct binding studies of phospholipase A_2 (or specifically modified derivatives) with organized lipid-water interfaces using spectroscopic techniques indicated that Trp^3 and Tyr^{69} are part of the IRS (22). Protection experiments against specific tryptic cleavage of the Arg^6-Ser^7 linkage were performed in the presence of micellar substrate analogs and pointed to a possible role of Arg^6 in the IRS. Upon selective carboxymethylation of Met^8, the enzyme is no longer able to bind lipid-water interfaces (27). Equilibrium gel filtration and ultraviolet difference spectroscopy have been used to study the interaction process as a function of pH. Anchoring of the enzymes into micelles of substrate analogs appears to be governed by the protonation state of a single group with pK = 8.3. This residue has been identified as the α-NH_3^+ group of the N-terminal amino acid Ala^1 (21-22). In summary these results strongly suggest that the IRS embraces at least the rather hydrophobic N-terminal sequence Ala^1.Leu.Trp.Gln.Phe.Arg.Ser.Met^8 and Tyr^{69}. In the 3D-structure of the zymogen this latter residue is located close to the N-terminus. (*Cf.* Figures 10A, B).

The particular role of the N-terminal amino acid Ala^1 of phospholipase A_2 in the recognition process of lipid-water interfaces has been demonstrated using a number of phospholipase A_2 analogs in which the polypeptide chain has been elongated, shortened or in which native L-Ala^1 has been replaced by other amino acids (28-29). A number of these modified enzymes are compiled in Figure 7. Both

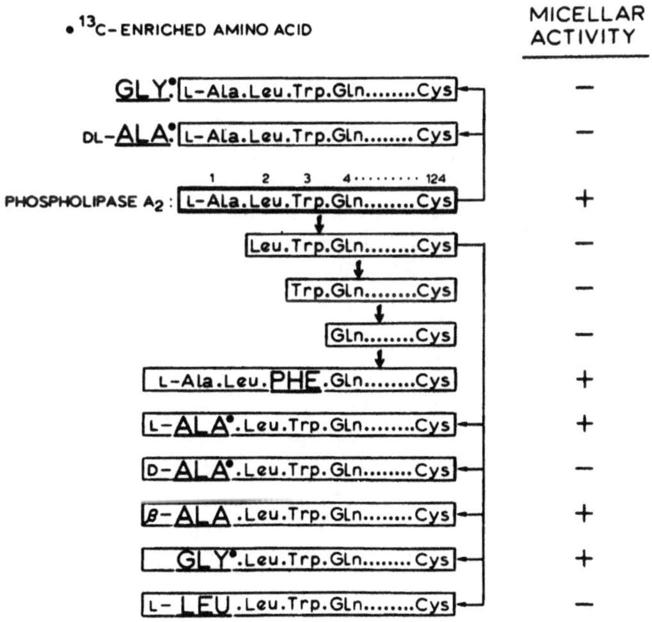

Fig. 7. *Scheme depicting selective substitution and addition of amino acids at the N-terminus of phospholipase A2.* ✱ *stands for 90%* ^{13}C *enriched amino acid (3-*^{13}C *D- and L-Ala; 1-*^{13}C *Gly). For experimental details see Slotboom and de Haas (28) and Slotboom* <u>*et al.*</u> *(29).*

chain elongation with an additional amino acid (Gly, DL-Ala) and chain shortening yield proteins which have lost their activity toward micellar substrates. A few substitutions in the native polypeptide chain are not lethal to a functionally active IRS: Gly^1 or β-Ala^1 in stead of L-Ala^1 and Phe^3 in stead of Trp^3. However, the presence of the more bulky L-Leu^1 and even of the stereoisomeric D-Ala^1 as N-terminal gives rise to a polypeptide chain which no longer recognizes micellar substrates. It has to be remarked that in these chemically modified phospholipase A_2 analogs the active center around His^{48} is still intact and enzyme activity toward monomeric substrates is similar to that of the native phospholipase A_2. These results strongly suggest that the induction of the IRS requires a very exact juxtaposition of the α-NH_3^+ group and its negatively charged countnerpartner. Most probably the above modified phospholipases A_2 which have lost their activity toward micellar substrates are not longer able to bind to these lipid-water interfaces because the internal saltbridge is impossible. An interesting consequence of the fact that D-Ala^1-phospholipase A_2 has lost its affinity for micellar substrates is the possibility to resolve the mixture of diastereoisomers D-Ala^1 -phospholipase A_2+L-Ala^1-phospholipase A_2 by "affinity chromatography" (29). As shown in figure 8^A, DL-Ala^1-phospholipase A_2 elutes from Sephadex G_{100} at pH 6.0 as a single symmetrical peak in a position corresponding to its molecular weight of ≈14000. A simular elution experiment at pH 6.0 in which the column was loaded with DL-Ala^1-phospholipase A_2 and micelles of the substrate analog n-hexadecylphosphorylcholine is shown in Figure 8B. The large lipid micelles elute at the breakthrough volume of the column, but now two equal protein peaks are observed. The L- Ala^1-phospholipase A_2 elutes with the micelles at the void volume, whereas the D-Ala^1-phospholipase A_2, which has no affinity for the lipid micelles, elutes at the same position as under 8A. Because of the fact that the binding of phospholipase A_2 to lipid micelles is governed by the pK of 8.3 of the α-NH_3^+ group, the pure lipid- free L-Ala^1-phospholipase A_2 can be isolated from the void volume fraction in Figure 8B by a repeated elution over the same column now at pH 9.5 (Figure 8C). It has to be remarked that the success of the above described separation technique of DL-Ala^1-phospholipase A_2 into the pure D-&L-components is due to the peculiar conformation of the N-terminal L-Ala^1 in native phospholipase A_2, forming a salt-bridge with a buried carboxylate group.
Similar experiments using DL-Ala^{-1}-phospholipase A_2* do not result in any separation.

*DL-Ala^{-1}-phospholipase A_2 refers to phospholipase A_2 to which DL-Ala has been covalently linked to the N-terminal amino acid L-Ala^1 of the native enzyme.

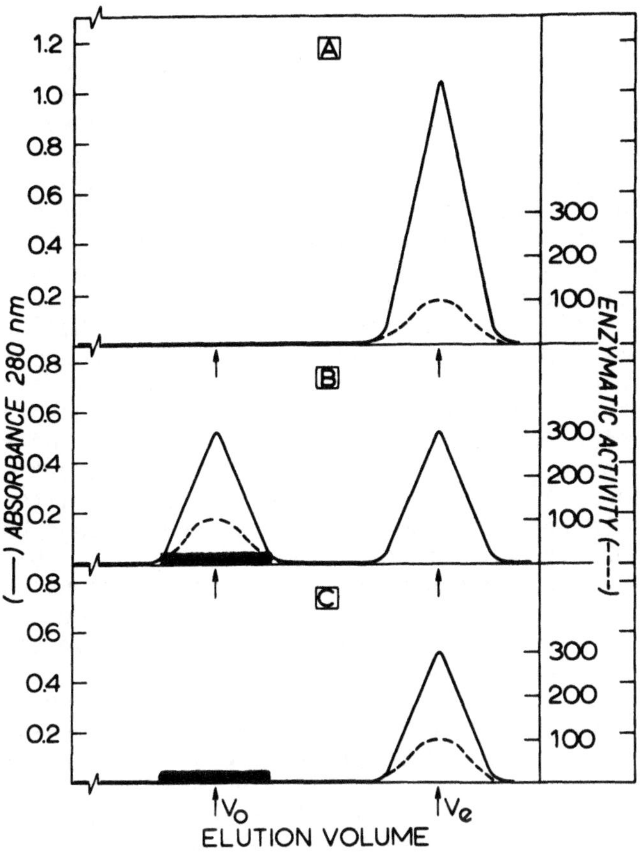

Fig. 8. *Schematic presentation of elution patterns of semisynthetic [DL-(3-^{13}C)-Ala$_1$]-phospholipase A_2 on a Sephadex G-100 column (60 x 1.2 cm). Experimental conditions:* A, *4 mg of [DL-(3-^{13}C)-Ala$_1$]-phospholipase A_2, pH 6.0, 0.1 M acetate, 0.1 M NaCl;* B, *4 mg of [DL-(3-^{13}C)-Ala$_1$]-phospholipase A_2 in the presence of 20 mg of n-hexadecylphosphorylcholine, pH 6.0, 0.1 M acetate, 0.1 M NaCl;* C, *fraction eluted at void volume (V_o) under B, pH 9.5, 0.1 M Tris-HCl, 0.1 M NaCl, 0.01 M EDTA. V_o = void volume of the column; V_e = elution volume of phospholipase A_2 (M. W. 14,000); ▬, position where micelles of n-hexadecylphosphorylcholine elute. Enzymatic activities (expressed in μequiv/ml) were determined as described by Nieuwenhuizen et al. (5) using a 1.5 fold higher sodium deoxycholate concentration. (From Slotboom et al. 29.)*

Role of Ca^{2+} in the interaction process

As discussed above the pancreatic phospholipases A_2 possess one "high-affinity" binding site for Ca^{2+}, located in the active center close to His[48]. This Ca^{2+} ion is absolutely required to enable the enzyme to hydrolyze monomeric substrates. Ca^{2+}, however, plays a second important role in the interaction between enzyme and organized lipid-water interfaces at alkaline pH. As shown in Figure 8B the enzyme binds to micellar interfaces at neutral or slightly acidic pH, but the complex dissociates at alkaline pH (Fig. 8C). It has been found by various direct binding techniques such as ultraviolet difference spectroscopy (22), fluorescence measurements (22), and microcalorimetry (23) that the interaction of phospholipase A_2 with lipid-water interfaces at basic pH is restored when specifically Ca^{2+} ions are present and that the complex remains intact up to high pH values. This is in agreement with the finding that the enzyme possesses a second optimum at pH 9.0 in the presence of relatively high Ca^{2+} concentrations (16,25) in addition to the optimum at pH 6.0 found at low Ca^{2+} concentration (30). These results have been tentatively explained by assuming that pancreatic phospholipase A_2 possesses at alkaline pH a second "low-affinity" Ca^{2+} binding site. Upon Ca^{2+}binding the protein undergoes a conformational change and the slightly altered polypeptide chain is able to anchor the lipid-water interface. Because micelle-binding requires an intact IRS, which is dependent on the internal salt bridge between the α-NH_3^+ group and a buried carboxylate, the effect of Ca^{2+} ions on the pK of the α-NH_3^+ group was studied. This was done by measuring the chemical shift of the resonance of phospholipase A_2 and some analogs containing a ^{13}C-enriched N-terminal amino acid as a function of pH. From the results given in Table 1, it can be seen that the pK of the α-NH_3^+ group of native phospholipase A_2 in the absence of Ca^{2+} ions is 8.4, which is in agreement with values determined previously by other methods. Upon addition of 50 mM Ca^{2+} the pK of the α-NH_3^+ group shifts to 9.0 and even to 9.3 when the Ca^{2+} concentration is increased to 100 mM. This increase in pK is rather specific for the N-terminal L-Ala[1] of native phospholipase A_2 and for Ca^{2+}ions. Mg^{2+} or Na^+ ions do not show any effect on this pK value, whereas the pK values of the α-NH_3^+ group of some enzymatically inactive phospholipase A_2 analogs having D-Ala as N-terminal amino acid or containing an extra amino acid (DL-Ala or Gly at the N-terminus show a drop from 8.3 to 7.6 upon addition of Ca^{2+}. These experiments strongly suggest that upon Ca^{2+} binding the salt bridge in native phospholipase A_2 moves to a much more hydrophobic surrounding. Due to this effect, which is even more pronounced in the presence of lipid-water interfaces, the salt bridge between the α-NH_3^+ group and the carboxylate, which triggers the IRS, remains intact up to higher pH values.

As regards the location of the "low-affinity"Ca^{2+} binding site in phospholipase A_2, no definite proposal can be given.

Table 1. pK VALUES OF THE N-TERMINAL α-NH_3^+ GROUP OF PORCINE PANCREATIC PHOSPHOLIPASE A_2 AND SOME ANALOGS

	Ca^{2+} (mM)			Mg^{2+} or NaCl (mM)
	0	50	100	100
1......124				
L-Ala*......Cys	8.4	9.0	9.3	8.4
D-Ala*......Cys DL-Ala*. L-Ala.......Cys Gly*. L-Ala.......Cys	8.3	7.6	-	7.8

*Stands for 90% ^{13}C enriched amino acid (3-^{13}C D- and L-Ala; 1-^{13}C Gly). Data obtained from 90.5 mHz ^{13}C-NMR titration curves at 25°C (1-2 mM protein solution in 50 mM Hepes buffer containing 10% D_2O).

In the absence of lipid-water interfaces the dissociation constant has been estimated by various techniques to be about 20 mM at pH 7.5, which means that even high resolution X-ray of the enzyme most probably will not be able to reveal its position. On the other hand, the absence of this site in the zymogen (31) and the influence of Ca^{2+} binding on the pK of the α-NH_3^+ group in phospholipase A_2 suggest that the second site will be located close to the N-terminal of the polypeptide chain.

Is the enzyme-lipid interaction reversible?

The model shown in Figure 3 implies that the interaction of phospholipase A_2 with organized lipid-water interfaces is reversible. There are a few reports, however, which might cast some doubt on this assumption:

Dervichian and coworkers (32) and Rietsch et al. (33) reported that the desorption of lipolytic enzymes from a lipid monolayer at the air-water interface is extremely slow. Dutilh (25) also favoured a rather irreversible interaction of phospholipase A_2 with lipid-water interfaces and proposed an unfolding of the polypeptide chain at any interface. His argument is based mainly on elution profiles such as shown in Figure 8B, which would not be compatible with dissociation constants of 1 mM.

On the other hand, there are numerous experiments which can only be explained if the enzyme anchors the interface in a reversible way, for example:

- the determination of apparent dissociation constants between phospholipase A_2 and lipid micelles by ultraviolet difference spectroscopy (22), fluorescence (22) or microcalorimetry (23). The observed saturation characteristics are not compatible with an irreversible interaction.
- in kinetic studies dilution of the reaction mixture immediately results in lower enzymatic activities.
- rapid "desorption" of the enzyme from a micelle-enzyme complex upon a pH change from 8 to 9, as monitored by ultraviolet difference spectroscopy.
- the rapid <u>endo</u>thermic heat effect observed in the microcalorimeter upon addition of EDTA to an enzyme-Ca^{2+}-micelle complex at pH 9.0. This heat effect equals the <u>exo</u>thermic heat evolved in the formation of the E-Ca^{2+}-micelle complex in the absence of EDTA (23).

The reversibility of the anchoring process is observed not only with the somewhat artificial <u>micellar</u> substrate analogs, but also in the interaction between enzyme and liposomes of long-chain phospholipids. As was demonstrated by Op Den Kamp <u>et al</u>. (34), the pancreatic phospholipase A_2 can anchor to the closely packed bilayers of long-chain lecithins only at the transition temperature. Liposomes of dimyristoyl lecithin are hydrolyzed rapidly at 23°C, but below <u>and above</u> this temperature no hydrolysis is observed. Apparently the enzyme is not able to penetrate into the pure gel or liquid crystalline phase.

The reversibility of the penetration process is shown in Figure 9. 1,2-Di-tetradecyl-rac-glycero-3-phosphorylcholine, a non-hydrolyzable ether analog of dimyristoyl lecithin has a transition temperature as determined by Differential Scanning Calorimetry of 26.7°C. Ultraviolet difference spectroscopy of a mixture of vesicles of this phospholipid + phospholipase A_2 against lipid vesicles alone as a function of increasing temperature shows a sharp decrease of turbidity in the mixture at the transition temperature, indicating a change in bilayer structure caused by interaction with the enzyme (curve a). Above the transition temperature a homogeneous liquid-crystalline phase is formed, the enzyme is squeezed out and the "absorption" signal returns to the value found below the transition temperature. The curve is fully reversible and can be obtained also by cooling down the mixture from temperatures above the transition temperature. In the presence of the zymogen (curve b), which is unable to recognize the lipid-water interface, no change in absorption is obtained. It has to be remarked that at alkaline pH and in the <u>absence</u> of Ca^{2+} curve b is obtained also for the active phospholipase A_2.

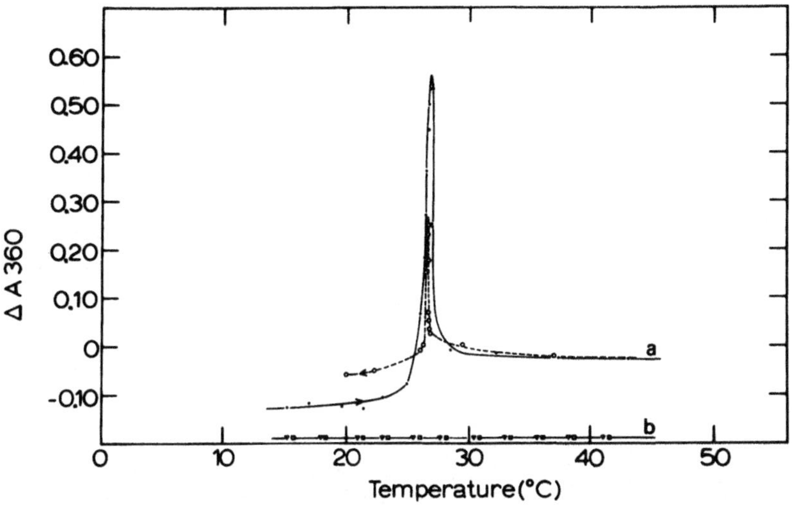

Fig. 9. *Change in turbidity at the transition temperature as measured by ultraviolet difference spectroscopy at 360 nm upon interaction of phospholipase A_2 with liposomes of 1,2-ditetradecyl-rac-glycero-3-phosphorylcholine at pH 8.5.* ●, *phospholipase A_2 in the presence of 25 mM Ca^{2+} (starting below the transition temperature);* o, *phospholipase A_2 in the presence of 25 mM Ca^{2+} (starting above the transition temperature);* Δ, *prophospholipase A_2 in the presence of 25 mM Ca^{2+};* □, *phospholipase A_2 in the absence of Ca^{2+}.*

This observation is in agreement with the results given in Table 1 and demonstrates once more that the enzyme can penetrate lipid-water interfaces at alkaline pH only after saturation of the second Ca^{2+} binding site.

Therefore, in our opinion there is no doubt that in general the interaction of a lipolytic enzyme with an organized lipid-water interface is reversible. This does not imply that the rate constants k_p and k_d in Figure 3 are of the same order of magnitude. Especially monomolecular surface films at the air-water interface may possess such a high interfacial energy (low surface pressure π) that proteins are nearly irreversibly adsorbed or even denatured (32,33). Therefore, kinetic studies of lipolytic enzymes using the monolayer technique should preferably be done at higher surface pressures where the substrate film possesses a low interfacial energy.

As regards elution profiles as shown in Figure 8B, they cannot be used as an indication for irreversible binding of the enzyme to the micelles.

The experimentally obtained dissociation constants between enzyme and lipid micelles (22) of about 1 mM are apparent values expressed

A

B

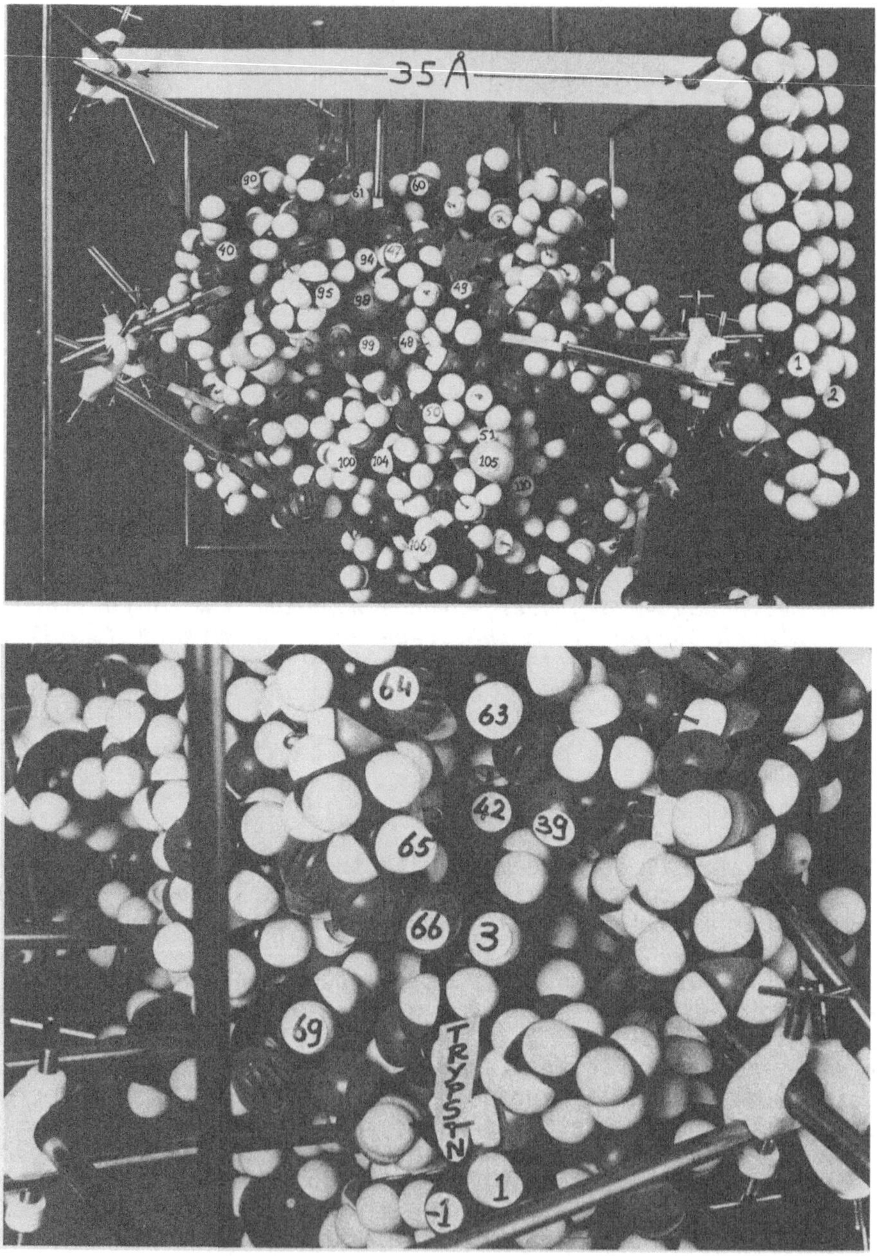

Fig. 10. *A survey of the active site region (fig. 10A) and of the N-terminus (fig. 10B) of the CPK molecular model of porcine pancreatic phospholipase* A_2 *(13).*

in the usual way as lipid monomers. It should be realized, however, that the enzymes do not bind free monomers but interact with a lipid aggregate. If one lipid micelle, consisting of about 200 monomers, binds and transports one enzyme molecule, the dissociation constant would not be 1 mM but 0.005 mM! In the latter case, elution profiles as shown in Figure 8B can be expected.

Mechanism of action of pancreatic phospholipase A_2

In this final part we will not try to unravel the mechanism through which the enzyme catalyzes the hydrolysis of monomeric substrates. This subject has been discussed recently by Sigman and Mooser (35) and Drenth *et al.* (13). What we would like to understand is how membrane-like structures such as organized lipid-water interfaces can control the activity of enzymes which by themselves are water-soluble. What are the parameters of the interface which control the activity of proteins anchored to the membrane?

The tremendous activation of phospholipase A_2 when acting upon certain organized lipid-water interfaces points to a very specific interaction between the enzyme and the aggregated substrate (17). Accelerating effects of 3 or 4 orders of magnitude compared with turnover numbers found in the hydrolysis of monomeric substrates cannot be explained by an increased substrate "concentration" only, and we feel that the particular lipid-water interface, dependent on its fine architecture, is able to induce, possibly in an "allosteric" way, certain changes in the active site of the enzyme.

The IRS or supersubstrate binding site (36) through which the enzyme anchors into the lipid phase has been shown to be not only functionally, but also topographically distinct from the active site. So one may wonder how the substrate monomers present in the interface arrive at the active center of the enzyme. The spatial separation between the active center around His^{48} and the IRS (N-terminal region of the polypeptide chain), suggested already by the two-dimensional structure given in Figure 2, can be seen also in Figures 10A and 10B, which are photographs of opposite sites of a space filling model of the zymogen of the porcine enzyme.

It might be argued, of course, that the X-ray structure given in Figure 10 represents the zymogen of porcine pancreatic phospholipase A_2 and not the active enzyme. However, even after splitting of the activation peptide by trypsin, it is very difficult to imagine how the newly formed N-terminal Ala^1 might form a salt bridge with a carboxylate in the neighbourhood of His^{48}. Bringing together of the N-terminal part of the chain with the active center would require a tremendous conformational change in the polypeptide chain. Taking into account the highly compact structure of this protein, caused by the presence of 7 disulfide bridges, such a conformational change seems to be highly improbable.

Acknowledgments

The authors gratefully acknowledge the collaboration of many colleagues during the past years. In particular they want to express their thanks to Drs. R.Verger, C.E.Dutilh and H.Meyer for their important contributions to the present study. Thanks are due to Mr. W.C. Puijk and Mr. R. Dijkman for their excellent technical assistance.
We are indebted to Drs. P.S. de Araujo and M.Y. Rosseneu for providing results prior to publication and to Mr. G.G. van Deijl (Laboratory of Molecular Biology, University of UTRECHT) for making the photographs of the molecular model.
The authors thank Drs. P.W.M. van Dijck for the determination of the transitiontemperature of 1,2-ditetradecyl-rac-glycero-3-phosphorylcholine.
This study was partly carried out under the auspices of the Netherlands Foundation of Chemical Research (S.O.N.), and with financial aid from the Netherlands Organization for the Advancement of Pure Research (Z.W.O.).
The 90.5 MHz ^{13}C-NMR measurements were carried out at the SON NMR facility in Groningen (The Netherlands).

References

1) L.L.M. van Deenen and G.H. de Haas, Adv. Lipid Res., 2 (1964) 167

2) P.P.M. Bonsen, G.H. de Haas and L.L.M. van Deenen, Biochim. Biophys. Acta, 270 (1972) 364.

3) W.A. Pieterson, J.J. Volwerk and G.H. de Haas, Biochemistry, 13 (1974) 1439.

4) R.D. Hershberg, G.H. Reed, A.J. Slotboom and G.H. de Haas, Biochemistry, 15 (1976) 2268, 3902.

5) W. Nieuwenhuizen, H. Kunze and G.H. de Haas, Methods Enzymol., 32B (1974) 147.

6) C.E. Dutilh, P.J. van Doren, F.E.A.M. Verheul and G.H. de Haas, Eur. J. Biochem., 53 (1975) 91.

7) A. Evenberg, H. Meyer, H.M. Verheij and G.H. de Haas, Biochim. Biophys. Acta, 491 (1977) 265.

8) C. Figarella, F. Clemente and O. Guy, Biochim. Biophys. Acta, 227 (1971) 213.

9) B. Arnesjö, J. Barrowman and B. Borgström, Acta Chem. Scand., 21 (1967) 2897.

10) W.C. Puijk, H.M. Verhey and G.H. de Haas, Biochim. Biophys. Acta, 492 (1977) 254.

11) E.A.M. Fleer, H.M. Verhey and G.H. de Haas, Eur. J. Biochem., (1977) in press.

12) A. Evenberg, H. Meyer, W. Gaastra, H.M. Verhey and G.H. de Haas, J. Biol. Chem. 252 (1977) 1189.

13) J. Drenth, C.M. Enzing, K.H. Kalk and J.C. A. Vessies, Nature, 264 (1976) 373.

14) J.P. Abita, M. Lazdunski, P.P.M. Bonsen, W.A. Pieterson and G.H. de Haas, Eur. J. Biochem., 30 (1972) 37.

15) L.H.M. Janssen, S.H. de Bruin and G.H. de Haas, Eur. J. Biochem., 28 (1972) 156.

16) R. Verger, M.C.E. Mieras and G.H. de Haas, J. Biol. Chem., 248 (1973) 4023.

17) R. Verger and G.H. de Haas, Ann. Rev. Biophys. Bioeng. 5 (1976) 77.

18) D. Chapman, Ann. N.Y. Acad.Sci., 137 (1966) 745.

19) D. Chapman, P. Byrne and G.G. Shipley, Proc. Roy. Soc., 290A (1966) 115.

20) S.J. Singer and G.L. Nicolson Science, 175 (1972) 720.

21) W.A. Pieterson, J.C. Vidal, J.J. Volwerk and G.H. de Haas, Biochemistry, 13 (1974) 1455.

22) M.C.E. van Dam-Mieras, A.J. Slotboom, W.A. Pieterson and G.H. de Haas, Biochemistry, 14 (1975) 5387.

23) P.S. de Araujo, M.Y. Rosseneu, A.J. Slotboom and G.H. de Haas, Manuscript in preparation.

24) J.J. Volwerk, W.A. Pieterson and G.H. de Haas, Biochemistry, 13 (1974) 1446.

25) C.E. Dutilh, Ph.D. Thesis, 1976, State University UTRECHT, The Netherlands.

26) M.Best-Belpomme and P. Dessen, Biochimie 55 (1973) 11.

27) F. van Wezel, A.J. Slotboom and G.H. de Haas, Biochim. Biophys. Acta, 452 (1976) 101.

28) A.J. Slotboom and G.H. de Haas, Biochemistry, 14 (1975) 5394.

29) A.J. Slotboom, M.C.E. van Dam-Mieras and G.H. de Haas, J. Biol. Chem., 252 (1977) 2948.

30) G.H. de Haas, P.P.M. Bonsen, W.A. Pieterson and L.L.M. van Deenen, Biochim. Biophys. Acta, 239 (1971) 252.

31) A.J. Slotboom, M.C.E. van Dam-Mieras, E.H.J.M. Jansen,F. Pattus, H.M. Verhey and G.H. de Haas,CNRS International Colloquium on Enzymes of Lipid Metabolism, April 1977, Strasbourg (France). Eds. P. Mandel, L. Freysz and S. Gatt, Plenum Press, New York, in press.

32) D.G. Dervichian, C. Préhu and J.P. Barque, C.R. Acad. Sci. Paris,276 (1973) 839.

33) J. Rietsch, F. Pattus, P. Desnuelle and R. Verger, J. Biol. Chem., 252 (1977) 4313.

34) J.A.F. Op Den Kamp, M. Th. Kauerz and L.L.M. van Deenen, Biochim. Biophys. Acta, 406 (1975) 169.

35) D.S. Sigman and G. Mooser, Ann. Rev. Biochem., 44 (1975) 913.

36) H. Brockerhoff and R.G. Jensen, Lipolytic Enzymes, (1974),p.17 New York: Academic Press.

Molecular Variation and Pathology

FAMILIAL LIPOPROTEIN DEFICIENCY

G. ASSMANN

KLINISCHE CHEMIE, UNIVERSITÄT KÖLN
5 KÖLN, JOSEPH-STELZMANN-STR.
WEST GERMANY

There are three genetically determined disorders in which one or more of the lipoprotein families are absent from plasma or their concentrations are extremely low: abetalipoproteinemia, hypobetalipoproteinemia, and Tangier disease (familial high density lipoprotein deficiency). Hypocholesterolemia is a common manifestation of all these diseases, which most easily can be distinguished on the basis of clinical symptoms and lipoprotein patterns.

Abetalipoproteinemia

Abetalipoproteinemia (Apo B deficiency) is a rare inborn error of lipid metabolism, characterized in the affected homozygote by the complete absence of chylomicrons, very low density lipoproteins (VLDL), and low density lipoproteins (LDL). The disease is associated with clinical evidence of fat malabsorption, retinitis pigmentosa, neuropathic ataxia, and acanthocytic red blood cells. The syndrom was first described in 195o in an 18-years-old-Jewish girl by Bassen and Kornzweig (Bassen-Kornzweig-syndrom). The characteristic lipoprotein changes in abetalipoproteinemia were discovered in 196o. As of today, approximately 35 cases have been reported. Comprehensive review articles on the clinical symptoms, the lipoprotein abnormality, the mode of inheritance, and the pathophysiology of this disease appeared in the literature.

Review articles

Schwartz,J.F.,Rowland,L.P.,Eder,H.,Marks,P.A.,Ossermann, E.F.,Hirschberg,E.,Anderson,H. (1963)
Bassen-Kornzweig syndrome: deficiency of serum ß-lipoprotein
Arch.Neurol. 8,438

Isselbacher,K.J.,Scheig,R.,Plotkin,G.R.,Caufield,J.B.(1964)
Congenital ß-lipoprotein deficiency: an hereditary disorder involving a defect in the absorption and transport of lipids.
Medicine 43,347

Wolff,O.H. (1965)
A-beta-lipoproteinemia
Ergebn.Inn.Med.Kinderheilk.23,19o

Farquhar,J.W.,Ways,P. (1966)
Abetalipoproteinemia
The Metabolic Basis of Inherited Disease, 2nd ed., J.B.Stanbury,J.B.Wyngaarden,D.S.Fredrickson,eds., McGraw-Hill,New York

Kahlke,W. (1967)
A-ß-lipoproteinemia
Lipids and Lipidoses, ed. by G.Schettler
Springer Verlag, New York

Fredrickson,D.S., Gotto,A.M., Levy,R.I. (1972)
Familial lipoprotein deficiency
The Metabolic Basis of Inherited Disease, 3rd ed., J.B.Stanbury,J.B.Wyngaarden,D.S.Fredrickson,eds., McGraw-Hill,New York

Kayden,H.J. (1972)
Abetalipoproteinemia
Amer.Rev.Med.23, 285

Kostner,G.(1976)
ApoB-deficiency (Abetalipoproteinemia): A model for studying the lipoprotein metabolism
Lipid Absorption: Biochemical and clinical aspects
K.Rommel,H.Goebell,R.Böhmer, eds.,MTP Press

Herbert,P.N., Fredrickson,D.S.(1976)
The Hypobetalipoproteinemias
Fettstoffwechsel,Handbuch der inneren Medizin.Fünfte Auflage,Band 7,Stoffwechselkrankheiten, 4. Teil
G.Schettler,H.Greten,G.Schlierf,D.Seidel, eds., Springer-Verlag,New York

lipoprotein abnormalities in abetalipoproteinemia

Levy,R.I.,Fredrickson,D.S.,Laster,L. (1966)
The lipoproteins and lipid transport in abetalipoproteinemia
J.Clin.Invest. 45, 531

Jones,J.W., Ways,P. (1967)
Abnormalities of high density lipoproteins in abetalipoproteinemia.
J.Clin.Invest. 46, 1151

Lees,R.S. (1967)
Immunological evidence for the presence of B protein (apoprotein of ß-lipoprotein) in normal and abetalipoproteinemic plasma
J.Lipid Res. 8, 396

Lees,R.S., Ahrens,E.H.,Jr. (1969)
Fat transport in abetalipoproteinemia
New Engl. J. Med. 28o, 1261

Gotto,A.M., Levy,R.I., John,K., Fredrickson,D.S. (1971)
On the nature of the protein defect in abetalipoproteinemia
New Engl. J. Med. 284, 813 (1971)

Forte,T., Nichols,A.V. (1972)
Application of electron microscopy to the study of plasma lipoprotein structure
Advanc.Lipid Res. 1o, 1

Kostner,G., Holasek,A., Bohlmann,H.G., Thiede,H. (1974)
Investigation of serum lipoproteins and apoproteins in abetalipoproteinemia
Clin.Sci.molec.Med. 46, 457

Scanu,A.M., Aggerbeck,L.P., Kruski,A.W., Lim,C.T. Kayden,H.J. (1974)
A study of the abnormal lipoproteins in abetalipoproteinemia
J.Clin.Invest. 53, 44o

Hypobetalipoproteinemia

Familial hypobetalipoproteinemia is considered a genetic disease distinct from abetalipoproteinemia. LDL are abnormally low but present; the disease appears to be transmitted as an autosomal dominant trait. The diagnostic criteria include the detection of a similar pattern in a first-degree relative, and the absence of a disease to which hypobetalipoproteinemia may be secondary. Moderate hypocholesterolemia and hypotriglyceridemia without acanthocytosis, fat malabsorption, and neuromuscular disease occur in most patients studied so far. The homozygous form of familial hypobetalipoproteinemia is clinically and chemically indistinguishable from abeta-lipoproteinemia.

case reports

Salt,H.B., Wolff,O.H., Lloyd,J.K., Fosbrooke,A.S., Cameron,A.H., Hubble,D.V. (196o)
On having no beta lipoprotein. A syndrome comprising a-beta-lipoproteinemia, acanthocytosis, and steatorrhea
Lancet II, 325

Van Buchem,F.S.P., Pol,G., de Gier,J., Böttcher,C.J.F., Pries,C. (1966)
Congenital ß-lipoprotein deficiency
Amer.J.Med. 4o, 794

Mars,H., Lewis,L.A., Robertson,A.L.,Jr., Butkus,A., Williams,G.H.,Jr. (1969)
Familial hypo-ß-lipoproteinemia. A genetic disorder of lipid metabolism with nervous system involvement
Amer.J.Med. 46, 886

Richet,G., Durepaire,H., Hartmann,L., Ollier,M.P., Polonovski,J., Maitrot,B. (1969)
Hypolipoproteinemie familiale asymptomatique predominant sur les beta-lipoproteines
Presse med. 77, 2o45

Levy,R.I., Langer,T., Gotto,A.M., Fredrickson,D.S. (197o)
Familial hypobetalipoproteinemia, a defect in lipoprotein synthesis.
Clin.Res. 18, 539

Mawatari,S., Iwashita,H., Kuroiwa,Y. (1971)
Familial hypo-ß-lipoproteinemia
J.neuro.Sci. 16, 93

Fosbrooke,A., Choksey,S., Wharton,B. (1973)
Familial hypo-ß-lipoproteinemia
Arch.Dis.Childh. 48, 729

Brown,B.J., Lewis,L.A., Mercer,R.D. (1974)
Familial hypobetalipoproteinemia: report of a case with psycho-motor retardation
Pediatrics 54, 11

Aggerbeck,L.P., McMahon,J.P., Scanu,A (1974)
Hypobetalipoproteinemia: Clinical and biochemical description of a new kindred with Friedreich`s ataxia
Neurology 24, 1o51

Cottrill,C., Glueck,C.J., Leuba,V., Millett,F., Puppione,D., Brown,W.V. (1974)
Familial homozygous hypobetalipoproteinemia
Metabolism 23, 779

Biemer,J.J., McCammon,R.E. (1975)
The genetic relationship of abetalipoproteinemia and hypobetalipoproteinemia: a report of the occurence of both diseases within the same family
J.Lab.clin.Med. 85, 556

Naito,H.K., Lewis,L.A. (1975)
Serum lipoproteins and lipids of familial hyper-,hypo-, or normo-ß-lipoproteinemic subjects during pregnancy
Clin.Chem. 21, 99o

Tamir,I., Levtow,O., Lotan,D., Lequin,C., Heldenberg,D., Werbin,B. (1976)
Further observations on familial hypobetalipoproteinemia
Clin.Genet. 9, 149

Glueck,C.J., Tsang,R.C., Mellies,M.J., Fallat,R.W., Steiner,P.M. (1976)
Neonatal familial hypobeta-lipoproteinemia
Metabolism 25, 611

Glueck,C.J., Gartside,P., Fallat,R.W., Sielski,J., Steiner,P.M. (1976)
Longevity syndromes: familial hypobeta- and familial hyperalphalipoproteinemia
J.Lab.Clin.Med. 88, 941

Tangier disease

Familial hypoalphalipoproteinemia (Tangier disease) was discovered in 196o by Fredrickson et al. and given its name from Tangier island which lies off the coast of Virginia in Chesapeake Bay and is the home of the original patients. To date, approximately 25 patients have been discovered. The clinical manifestations include: low plasma cholesterol and normal or elevated triglyceride concentrations; absence of the usual high density lipoproteins and abnormal composition of other lipoproteins in plasma; distinctive enlargement and coloration of the tonsils; frequently peripheral neuropathy; cholesteryl ester deposition in tissues accompanied by enlargement of spleen,liver,or lymph nodes, alterations in the cornea, intestinal mucosa, and other tissues.

Review articles

Fredrickson,D.S., Altrocchi,P.H. (1962)
Tangier disease (Familial cholesterolosis with high-density lipoprotein deficiency
Cerebral Sphingolipidoses,A symposium on Tay Sachs Disease and allied disorders, S.M.Aronson,B.W.Volk,eds.
Academic Press, New York

Fredrickson,D.S. (1966)
Familial high-density lipoprotein deficiency: Tangier disease
The Metabolic Basis of Inherited Disease, IInd. ed.
J.B.Stanbury,J.B.Wyngaarden,D.S.Fredrickson,eds.
McGraw-Hill, New York

Fredrickson,D.S.,Gotto,A.M., Levy,R.I. (1972)
Familial lipoprotein deficiency
The Metabolic Basis of Inherited Disease, IIIrd. ed.
J.B.Stanbury,J.B.Wyngaarden,D.S.Fredrickson
McGraw-Hill, New York

Assmann,G. (1976)
Tangier Krankheit
Fettstoffwechsel. Handbuch der inneren Medizin. Fünfte Auflage. Band 7. Stoffwechselkrankheiten. 4.Teil
G.Schettler,H.Greten,G.Schlierf,D.Seidel, eds.
Springer Verlag, New York

case reports

Fredrickson,D.S., Altrocchi,P.H., Avioli,L.V., Goodman,D.S., Goodman,H.C. (1961)
Tangier disease. Combined clinical staff conference at the National Institutes of Health.
Ann.Intern.Med. 55, 1o16

Fredrickson,D.S., Young,O., Shiratori,T., Briggs,N.(1964)
The inheritance of high density lipoprotein deficiency (Tangier disease)
J.Clin.Invest. 43, 228

Hoffman,H.N., Fredrickson,D.S. (1965)
Tangier disease (familial high density lipoprotein deficiency). Clinical and genetic features in two adults.
Am.J.Med. 39, 582

Kocen,R.S., Lloyd,J.K., Lascelles,P.T., Fosbrook,A.S. Williams,D. (1967)
Familial alpha-lipoprotein deficiency (Tangier disease) with neurological abnormalities
Lancet I, 1341

Kummer,H., Laissue,J., Spiess,H., Pflugshaupt,R., Bucher,U (1968)
Familiäre Analphalipoproteinämie (Tangier Krankheit)
Schweiz.med.Wschr. 98, 4o6

Huth,K., Kracht,J., Schoenborn,W., Fuhrmann,W. (197o)
Tangier-Krankheit (Hyp-alpha-lipoproteinämie)
Dtsch.med.Wschr. 95, 2357

Clifton-Bligh,P., Nestel,P.J., Whyte,H.M. (1972)
Tangier disease: report of a case and studies of lipid metabolism
New Engl.J.Med. 286, 567

Greten,H., Hannemann,T., Gusek,W., Vivell,O. (1974)
Lipoproteins and lipolytic plasma enzymes in a case of Tangier disease
New Engl.J.Med. 291, 548

Brook,J.G., Lees,R.S., Yules,J., Cusack,B. (1977)
Tangier disease in New England-report of a case
JAMA (J.Am.Med.Assoc.) in press

lipoprotein abnormalities

Lux,S.E., Levy,R.I., Gotto,A.M., Fredrickson,D.S. (1972)
Studies on the protein defect in Tangier disease. Isolation and characterization of an abnormal high density lipoprotein
J.Clin.Invest. 51, 25o5

Kostner,G., Holasek,A., Schoenborn,W., Fuhrmann,W. (1972)
Immunchemische Untersuchung und isoelektrische Fokussierung des Serums eines Patienten mit Tangier Krankheit
Clin.chim.Acta 38, 155

Utermann,G., Menzel,H.J., Schoenborn,W. (1972)
Plasma lipoprotein abnormalities in a case of primary high-density lipoprotein (HDL) deficiency
Clin.Genetics 8, 258

Assmann,G., Smootz,E., Adler,K., Capurso,A., Oette,K. (1977)
The lipoprotein abnormality in Tangier disease. Quantitation of A apoproteins
J.Clin.Invest. 59, 565

Assmann,G., Herbert,P.N., Fredrickson,D.S., Forte,T. (1977)
Isolation and characterization of an abnormal high density lipoprotein in Tangier disease
J.Clin.Invest., in press

Assmann,G., Simantke,O., Schaefer,H.E., Smootz,E. (1977)
Characterization of high density lipoproteins in patients heterozygous for Tangier disease
J.Clin.Invest., in press

tissue abnormalities

Waldorf,D.S., Levy,R.I., Fredrickson,D.S. (1967)
Cutaneous cholesterol ester deposition in Tangier disease
Arch.Derm. 95, 161

Engel,W.K., Dorman,J.D., Levy,R.I., Fredrickson,D.S. (1967)
Neuropathy in Tangier disease. Alpha-lipoprotein deficiency manifesting as familial recurrent neuropathy and intestinal lipid storage
Arch.Neurol. (Chic.) 17, 1

Shacklady,M.M., Djardjouras,E.M., Lloyd.J.K.(1968)
Red-cell lipids in familial alpha-lipoprotein deficiency (Tangier disease)
Lancet II, 151

Spiess,H., Ludin,H.P., Kummer,H. (1969)
Polyneuropathie bei familiärer Analphalipoproteinämie
Nervenarzt 4o, 191

Kracht,J., Huth,K., Schoenborn,W., Fuhrmann,W. (197o)
Hypo-alpha-lipoproteinämie (Tangier Disease)
Verhdl.Dtsch.Ges.Path. 54, 355

Bale,P.M., Clifton-Bligh,P., Benjamin,B.N.P., Whyte,H.M. (1971)
Pathology of Tangier disease
J.Clin.Pathol. (Lond.) 24, 6o9

Kocen,R.S., King,R.H.M., Thomas,P.K., Haas,L.F. (1973)
Nerve biopsy findings in two cases of Tangier disease
Acta Neuropath. 26, 317

Ferrans,V.J., Fredrickson,D.S. (1975)
The pathology of Tangier disease
Amer.J.Pathol. 78, 1o1

Gheorghiu,Th., Assmann,G., Schaefer,H.E. (1976)
Endoscopic findings in Tangier disease.
Endoscopy 8, 164

Katz,S.S., Small,D.M., Brook,J.G., Lees,R.S. (1977)
The storage lipids in Tangier disease. A physical chemical study
J.Clin.Invest. 59, 1o45

FAMILIAL HYPERCHOLESTEROLAEMIA

G.R. THOMPSON

Medical Research Council Lipid Metabolism Unit

Hammersmith Hospital, London, England

Familial Hypercholesterolaemia (FH), or familial type II hyperlipoproteinaemia (1), is inherited in an autosomal dominant manner and is characterised by abnormally high levels of low density lipoprotein (LDL or β-lipoprotein) in plasma from birth. It has been estimated that FH affects 1:500 of the population of the United States of America and is responsible for 1:20 of all cases of myocardial infarction (2). Thus the disease is one of the commoner disorders inherited by humans and one which exerts a considerable toll in terms of premature morbidity and mortality. The current interest in attempting to reduce the prevalence of atherosclerosis in Western Countries has focussed attention on FH as a means of proving or disproving the lipid hypothesis of atherosclerosis. If the well-established relationship between hypercholesterolaemia and coronary heart disease is neither fortuitous nor due to some unknown third factor, then hypercholesterolaemia must either be a cause or a consequence of coronary heart disease. Since the hypercholesterolaemia of FH antedates the coronary heart disease by several years, it seems reasonable that the former causes the latter, and not vice versa. If so, effective reduction of hypercholesterolaemia in this monofactorial situation should result theoretically in regression of atherosclerosis, assuming that the atheromatous lesions in FH respond in an analagous manner to those which have been induced experimentally in monkeys and pigs (3, 4). Thus FH merits study not only as a disease entity in its own right but also because it provides a unique opportunity to determine whether lipid-lowering therapy is likely to be beneficial in the prevention and regression of premature coronary heart disease in the population at large (1). As Robert Frost once said: 'Nature is always hinting at us. It hints over and over again. And suddenly we take the hint'.

CLINICAL FEATURES

Homozygous inheritance of the FH gene, which is fortunately rare, is characterised by the appearance during childhood of cutaneous and tendinous xanthomata and the premature onset of aortic and coronary atheroma in association with markedly elevated levels of LDL in plasma. Sudden death before the age of 30 is the rule rather than the exception (5), and is attributable to acute coronary insufficiency secondary to coronary ostial stenosis (6). In life aortic involvement is manifested by an aortic systolic murmur without an ejection click, a gradient across the aortic valve and narrowing of the aortic root on aortography (7, 8). At post-mortem both the aortic valve and especially the supravalvar region of the aorta are infiltrated, thickened and narrowed by atheroma (9, 10). Hisologically the lesions closely resemble 'ordinary' atheroma but contain more cholesterol and more numerous foam cells. Atheroma is also found distally throughout the aorta, although to a lesser extent than proximally (11), and in the carotid, basilar and pulmonary arteries.

Heterozygotes usually present in early to mid-adult life with ischaemic symptoms such as angina, intermittent claudication or even vertebro-basilar insufficiency (12). Their degree of hypercholesterolaemia is less severe than in homozygotes and they do not exhibit the cutaneous planar xanthomata which are characteristic of the latter. The majority of heterozygotes have tendon xanthomata by the age of 30 (13), commonly affecting the Achilles tendons and the extensor tendons of the hands. Corneal arcus is also common but xanthelasma less so (14). FH is more often associated with a type IIa phenotype than with a type IIb, although the proportion of patients exhibiting the latter rises with age (13). Up to 30 % of heterozygotes have an aortic systolic murmur (15) but autopsy studies reveal only supravalvar aortic atheroma, without involvement of the valve itself (11). Coronary atheroma is common, with an ominous predilection for the main stem of the left coronary artery (16).

MOLECULAR BASIS: IN VITRO STUDIES

In 1973 Goldstein & Brown (17) reported that fibroblasts cultured from the skin of an FH homozygote had a fifty-fold higher level of HMG-CoA reductase activity, the rate-limiting enzyme of cholesterol synthesis, than normal fibroblasts. Unlike the latter the activity in FH cells was not inhibited by the addition of LDL to the culture medium. Subsequently they showed that labelled LDL was bound to normal fibroblasts by means of a specific high affinity, saturable mechanism which was lacking in cells from FH homozygotes (18). In a series of further studies, which they recently summarized (19), these authors postulated that high affinity binding of LDL was dependent upon the presence of specialized receptors on the cell surface of fibroblasts and that these LDL receptors were lacking to a greater or lesser extent in FH cells. Using this approach they subdivided homozygotes into receptor-negative or receptor-defective categories (20), whereas cells from heterozygotes showed only about a 50 % reduction in their complement of LDL receptors. Under normal circumstances the binding of LDL to these receptors initiates the following chain of events: cellular uptake of LDL; incorporation of LDL into lysosomes, with hydrolysis of both the cholesterol ester and apoB moieties of LDL; suppression of HMG-CoA reductase activity and thus of cholesterol synthesis in microsomes by newly-released free cholesterol, the latter subsequently being re-esterified as a result of the activation of the enzyme acyl-CoA:cholesterol acyltransferase (19). In receptor-negative cells absence of specific binding of LDL precludes any of the subsequent steps from occurring at a normal rate.

An alternative explanation for the defect in FH was proposed by Stein et al (21). These authors found no decrease in LDL binding in cultured fibroblasts from an FH homozygote but showed a marked decrease in the rate of 'internalization' or cellular uptake of bound LDL. However, Brown and Goldstein (2) subsequently claimed to have demonstrated the absence of LDL receptors in the same strain of cells, and suggested that the discrepancy was due to methodological differences.

The existence of LDL receptors has also been demonstrated in cells cultured from the intima and media of human foetal aorta (22) and in freshly isolated lymphocytes from normal subjects, but not from patients with homozygous FH; a 72 hour period of incubation in an LDL-deficient medium is necessary, however, to unmask these high affinity binding sites (23). Fogelman et al (24) had earlier shown that cholesterol synthesis in leucocytes was enhanced by prior incubation in lipoprotein-deficient serum and that this occurred to a greater extent than normal in cells from FH heterozygotes. Subsequent studies showed enhanced activity of HMG-CoA reductase in FH leucocytes, together with evidence of an increased rate of efflux of cholesterol into the incubation medium (25). These authors proposed that leakage of cholesterol from the cell is the primary defect in FH and that the increased activity of HMG-CoA reductase and consequent enhancement of cholesterol synthesis are secondary events.

The validity of the LDL receptor hypothesis in the context of FH has recently been examined by Myant (26). In his view one of its main attractions is that it explains the selectivity of the increase in LDL, as opposed to other cholesterol-carrying lipoproteins. Furthermore the complete or partial absence of LDL receptors is consistent with the most characteristic _in vivo_ abnormality, namely a reduced fractional rate of catabolism of LDL. The fact that catabolism occurs at all in receptor-negative homozygotes is explicable on the basis of non-specific uptake and degradation of LDL which proceeds at similar rates in normal and FH fibroblasts (27), assuming that the latter behave similarly to the cells involved in LDL catabolism. However, the receptor hypothesis is less easy to reconcile with the lack of evidence of any *in vivo* increase in cholesterol synthesis, since non-specific uptake of LDL into homozygous FH fibroblasts does not appear to inhibit their HMG-CoA reductase activity. But although much remains to be explained nevertheless the work of Goldstein and Brown has had a considerable impact on the field of lipoprotein metabolism and has stimulated others to utilize similar techniques.

MOLECULAR BASIS: *IN VIVO* STUDIES

Most of the earlier *in vivo* studies were aimed at measuring cholesterol turnover in FH. In most instances there was no evidence of any increase in cholesterol synthesis in FH patients, whether homozygous or heterozygous (26). An exception was the young homozygote reported by Bilheimer *et al.* (28) as having an increased rate of synthesis, although the absence of any data in control subjects of comparable age throws doubt on the significance of this observation. More recently, Bhattacharyya *et al.* (29) failed to show any increase in cholesterol synthesis in heterozygotes, in spite of their being on a nil or low intake of dietary cholesterol. Furthermore, their absorption of dietary cholesterol has been shown to be normal (30). Miettinen and Aro (31) reported that bile acid excretion was reduced in a family of heterozygotes whom they studied but other workers have found normal values in both homozygotes and heterozygotes, as well as a normal response to cholestyramine (32). Thus it would appear that the absolute rates of synthesis of both cholesterol and bile acids are usually normal in FH, although there is undoubtedly quite a marked reduction in the fractional rate of clearance of cholesterol from a considerably expanded rapidly-exchangeable pool (29).

In an attempt to answer the question as to how the rapidly exchangeable pool of cholesterol, which includes plasma, gets so enlarged, Langer *et al.* (33) measured the rate of turnover of the radioiodinated protein moiety of LDL, namely apoB, in a group of FH heterozygotes. Under steady-state conditions and using the technique of compartmental analysis introduced by Matthews (34), they found that FH patients differed from control subjects in having a lower fractional rate of catabolism (FCR) of LDL-apoB, even though the absolute rate of catabolism, and thus the synthetic rate, was within normal limits. Using a similar approach Packard *et al.* (35) confirmed the reduced FCR in FH heterozygotes but found that this was accompanied by a modest increase in LDL synthesis. Reichl *et al.* (36) and Simons *et al.* (37) showed that the FCR of LDL-apoB was even more reduced in homozygotes, and that this was accompanied by a marked increase in the rate of apoB synthesis. Recently, Thompson and his colleagues (38, 39) performed non-steady-state studies of LDL turnover in FH patients after having reduced the expanded LDL pool to near-normal size by means of plasma exchange. The results showed that the FCR remained remarkably constant, irrespective of the changing size of the LDL pool, and suggested that the reduced FCR is an intrinsic feature of the disease rather than a secondary phenomenon. This conclusion is strengthened by the normal FCR of LDL in patients with primary hypobetalipoproteinaemia, in whom LDL pool size is greatly reduced due to decreased apoB synthesis (40). Thus it would seem that the hypercholesterolaemia of FH can be at least partly explained on the basis of defective catabolism of LDL, the major carrier of cholesterol in plasma.

The mechanism of the increased rate of synthesis of apoB in homozygotes has also been investigated recently (41). The results suggest that more LDL-apoB is secreted into the plasma of these subjects than can be explained on the basis of their normal rate of synthesis and conversion of very low density lipoprotein (VLDL) to LDL. This is in keeping with the concept of Brown & Goldstein (2) that the function of the LDL receptor in liver cells is to regulate the rate of synthesis of apoB, and that absence of such receptors, as in homozygotes, can lead to the direct secretion of LDL from the liver into plasma.

The actual site of catabolism of LDL *in vivo* remains controversial. The somewhat unphysiological studies of Sniderman *et al.* (42) in hepatectomised pigs suggest that LDL catabolism takes place mainly in extrahepatic tissues, whereas the fact that cholestyramine increases the FCR of LDL as well as enhancing bile acid excretion points to the liver as a major site of catabolism of both the protein and lipid components of LDL (43). Experimental data suggests that hepatic tissues undoubtedly have the ability to bind LDL, both *in vitro* (44) and *in vivo* (45), in an analogous manner to that seen in fibroblasts, but without quite the same degree of specificity.

Analyses of the amino acid composition of LDL-apoB have consistently failed to show any abnormality in FH (46, 47). However, undoubted differences exist between the lipid composition of LDL from FH patients and normal subjects, the former having a lower proportion of triglyceride than the latter (48). Of perhaps greater interest is the increased cholesterol:protein ratio of LDL (49), which mainly reflects an increase in free cholesterol (50). This is accompanied by a decrease in the ratio of lecithin:sphingomyelin (51, 52). These abnormalities are more marked in homozygotes than in heterozygotes but become less abnormal following plasma exchange (53). This suggests that the increased content of sphingomyelin and cholesterol are causally related and that they are probably secondary to the prolonged half-life of LDL in the circulation.

In this context it is of interest that fibroblasts from FH homozygotes also have an increased content of sphingomyelin, which diminishes after they have been incubated in a lipoprotein-free medium (54). It seems unlikely that the abnormalities of LDL composition are primary phenomena since analogous changes have been observed in animals made hyperbetalipoproteinaemic by methods which decrease LDL catabolism (51, 55).

MODEL FOR REGRESSION OF ATHEROSCLEROSIS

The premature onset of atherosclerosis is an inevitable accompaniment of FH. This tendency is particularly marked in homozygotes, many of whom die suddenly in childhood or in their teens (5). Few, if any, have survived beyong the age of 30 (1). The considerable risks of developing coronary and peripheral vascular disease in heterozygotes are also well documented, and the averaged results of three surveys are shown in the Table. It is clear that males are affected at an earlier age than females, although this sex advantage is apparently lost in females who smoke cigarettes (14). Slack (56) found that male heterozygotes have a 23 % expectation of death from coronary heart disease before the age of 50. Heiberg (15, 57) studied several Norwegian families with FH and found that half the males were dead by the age of 66, whereas half the females could expect to live to the age of 75. In his series both the age of onset of ischaemic symptoms and the appearance of xanthomata were related to the height of the serum cholesterol. Few patients with serum cholesterol levels below 10 mmol/l developed xanthomata. But half the males and half the females who did develop xanthomata were dead within 8 to 12 years respectively of their first appearance.

TABLE

Risks of Coronary Heart Disease in FH. Pooled data from Slack (56), Stone et al (65) and Beaumont et al (14)*

Age	Males	Females
30	5 %	0 %
40	29 %	20 %
50	60 %	31 %
60	69 %	45 %

* Reference (14) included patients with Peripheral Vascular Disease.

During the past 10 years considerable efforts have been made to reduce the premature morbidity and mortality from coronary heart disease in FH by reducing the severity of the hypercholesterolaemia. The mainstay of treatment has been the anion-exchange resin cholestyramine which specifically lowers LDL levels. At the present time a large-scale trial of the efficacy of this drug in the primary prevention of coronary heart disease in patients with type II hyperlipoproteinaemia (some of whom have FH) is being carried out in the United States, but the results will not be available for sometime. Another proposed trial, but of secondary intervention, has recently been described by Buchwald et al (58). This is based on the efficacy of ileal-bypass as a means of lowering the serum cholesterol and involves the use of coronary arteriography as a means of assessing progression or regression of established coronary heart disease in patients with hypercholesterolaemia, some of whom will presumably have FH.

Ileal bypass is undoubtedly as effective as cholestyramine in the management of heterozygous FH but neither form of treatment adequately controls the hypercholesterolaemia of homozygotes. An alternative approach was conceived by Starzl et al (59) who performed a portacaval shunt in a young female homozygote; the ensuing reduction in hypercholesterolaemia was said to be accompanied by evidence of regression of her atheroma (60). Subsequent experience with this operation has been less encouraging, however, since on average it results in only an approximately 30 % reduction in serum cholesterol levels (61). The mechanism whereby portacaval shunts influence lipoprotein metabolism remains controversial but a decrease in hepatic cholesterol synthesis has been documented both in animals (62) and in a patient with homozygous FH (28).

Because of the difficulty of controlling the hypercholesterolaemia of homozygous and severe heterozygous FH by conventional means, Thompson et al (63) introduced the use of plasma exchange. This is performed with a continuous flow blood cell separator (Aminco "Celltrifuge"), as illustrated in Fig.1. Heparinised venous blood from the patient is continuously separated into plasma and cellular components by centrifugation. By means of an ingenious seal and peristaltic pumps the plasma is collected for disposal, whereas the cellular components are returned to the patient via a vein in the opposite arm, after being reconstituted with plasma protein fraction (PPF). The latter is a cholesterol-free solution containing mainly human serum albumin. In general 3-4 litres of plasma are exchanged on each occasion at whole blood flow rates of up to 70 ml per minute, the entire procedure lasting about 2 hours.

The rationale for plasma exchange is that the hypercholesterolaemia of FH is largely due to defective catabolism of LDL, as illustrated by Fig.2. Thus physical removal of LDL, which is mainly located intravascularly (32), is a logical approach to the

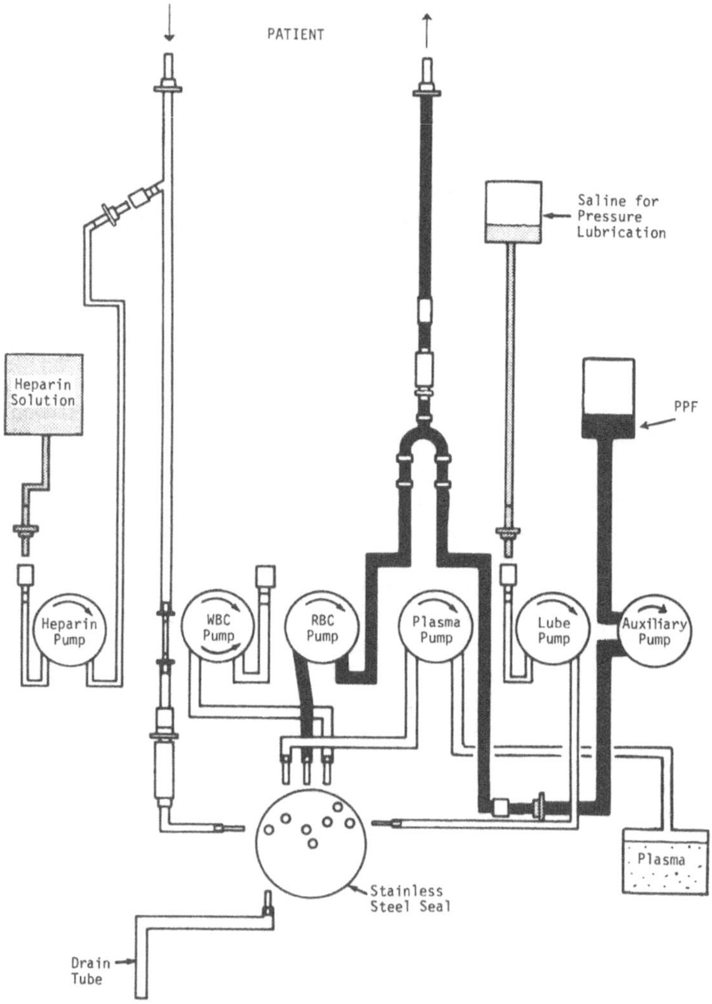

Fig.1. Simplified circuit diagram of "Celltrifuge". The stainless steel seal on top of the centrifuge bowl is continuously lubricated with sterile saline.

problem. Immediately after plasma exchange the serum cholesterol level decreases to one-third of the pre-exchange value, and then gradually rises at the rate of about 5 mmol/l per week in homozygotes and 2.5 mmol/l per week in heterozygotes, the rate of rise being slowed to some extent by hypocholesterolaemic agents. However, even so, effective control of hypercholesterolaemia necessitates weekly plasma exchange in all homozygotes and also in some heterozygotes, although in others plasma exchange every 2 weeks in probably sufficient.

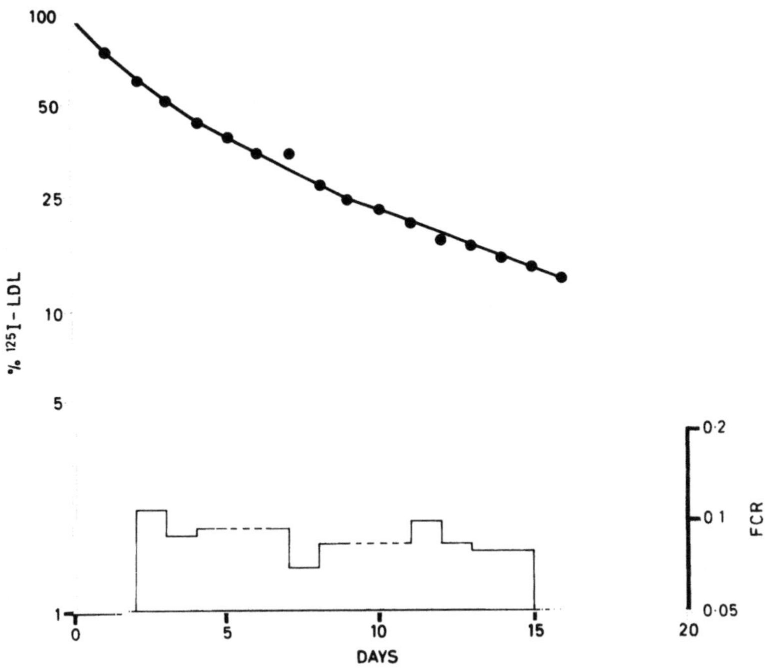

Fig.2. Slow turnover of ^{125}I-LDL in patient with homozygous FH, manifested by the prolonged half-life of LDL in plasma (7.4 days) and reduction of the daily fractional catabolic rate (FCR) to less than 0.1.

Each patient selected for a course of plasma exchange at the Hammersmith Hospital first undergoes coronary angiography and aortography, in order to delineate the extent of atheroma in these vessels. These investigations are repeated after a period of approximately 2 years of plasma exchange, so as to obtain an objective assessment of each patient's progress. Conventional drug therapy with β-blockers for angina and hypocholesterolaemic agents (usually cholestyramine plus either nicotinic acid or clofibrate) is maintained throughout. To date 4 homozygotes and 3 heterozygotes have been or are regularly undergoing plasma exchange. Two patients have died; one of them, a homozygous female, died following an aorto-coronary bypass, the other, a male heterozygote, from a myocardial infarct. Neither death was attributable to plasma exchange. Of the other five patients two are male homozygotes who remain asymptomatic and in whom cutaneous and tendon xanthomata show signs of resolution. Plasma exchange was discontinued in the fourth homozygote because the smallness of her veins precluded good flow rates; subsequently she underwent a porta-caval shunt. Two male heterozygotes remain well and have had some amelioration of their angina. In one of them weekly plasma exchange maintains serum cholesterol levels consistently within the normal range (< 6.5 mmol/l).

There have been no untoward reactions during the last 100 plasma exchanges, in particular there has been no evidence of sensitisation to PPF, no haemorrhage or sepsis and no cardiovascular complications. All patients have remained Australia antigen negative. Episodes of hypocalcaemia and hypokalaemia have been prevented by the routine addition of 1.5 mmol potassium (as potassium chloride) and 0.45 mmol calcium (as calcium gluconate) to each 400 ml of PPF. Using this approach serum electrolytes remain normal throughout. The platelet count decreases by up to 50 % by the end of the procedure but returns to normal within one week. Thus plasma exchange at weekly or fortnightly intervals has proved to be a safe and practicable procedure which results in substantially better control of hypercholesterolaemia than could otherwise be achieved. Whether this will result in demonstrable improvement of coronary atheroma remains to be seen. However, the recent report of Barndt et al (64) is encouraging, in that it showed angiographic evidence of regression of femoral atheroma in FH heterozygotes in whom adequate control of hyperlipoproteinaemia had been achieved by conventional means.

REFERENCES

1. Fredrickson, D.S.,and R.I. Levy, in The Metabolic Basis of Inherited Disease, Third Edition. (Eds) J.B. Stanbury, J.B. Wyngaarden and D.S. Fredrickson, McGraw Hill, New York (1972) pp. 545-614.
2. Brown, M.S., and J.L. Goldstein. New Engl. J. Med. 294: 1386 (1976).
3. Armstrong, M.L., E.D. Warner, and W.E. Connor. Circulation Res. XXVII:59 (1970).
4. Daoud, A., J. Jarmolych, J.M. Augustyn, K.E. Fritz, J.K. Singh, and K.T. Lee. Arch. Pathol. Lab. Med. 100:372 (1976).
5. Cook, C.D., H.L. Smith, C.W. Geissen, and G.L. Berdez. Amer. J. Dis. Child. 73:326 (1947).
6. Rigadon, R.H., and G. Willeford. J. Amer. Med. Ass. 142: 1268 (1950).
7. Stanley, P., C. Chartrand, and A. Davignon. New Engl. J. Med. 273:1378 (1965).
8. Bialostozky, D., M.Luengo, C. Magos, and E. Zorrilla. Amer. J. Cardiol. 36:509 (1975).
9. Barr, D.P., S. Rothbard, and H.A. Eder. J. Amer. Med. Ass. 156:943 (1954).
10. Wennevold, A., and J.G. Jacobsen, Amer. J. Med. 50:823 (1971).
11. Roberts, W.C., V.J. Ferrans, R.I. Levy, and D.S. Fredrickson. Amer. J. Cardiol. 31:557 (1973).

12. de Gennes, J.L., J. Rouffy, and F. Chain. Bull. Soc. Med. Hop. Paris 119:569 (1968).
13. Kwiterovich, P.O., D.S. Fredrickson, and R.I. Levy. J. Clin. Invest. 53: 1237 (1974).
14. Beaumont, V., B. Jacotot, and J.-L. Beaumont. Atherosclerosis 24:441 (1976).
15. Heiberg, A. Acta med. Scand. 198:249 (1975).
16. Bloch, A., R.E. Dinsmore, and R.S. Lees. Lancet i:928 (1976).
17. Goldstein, J.L., and M.S. Brown. Proc. Nat. Acad. Sci. U.S. 70:2804 (1973).
18. Brown, M.S., and J.L. Goldstein. Proc. Nat. Acad. Sci. U.S. 71:788 (1974).
19. Brown, M.S., and J.L. Goldstein. Science 191:150 (1976).
20. Goldstein, J.L., S.E. Dana, G.Y. Brunschede, and M.S. Brown. Proc. Nat. Acad. Sci. U.S. 72:1092 (1975).
21. Stein, O., D.B. Weinstein, Y. Stein, and D. Steinberg. Proc. Nat. Acad. Sci. U.S. 73:14 (1976).
22. Goldstein, J.L., and M.S. Brown. Arch. Pathol. 99:181 (1975).
23. Ho, Y.K., M.S. Brown, D.W. Bilheimer, and J.L. Goldstein. J. Clin. Invest. 58:1465 (1976).
24. Fogelman, A.M., J. Edmond, A. Polito, and G. Popják. J. Biol. Chem. 248:6928 (1973).
25. Fogelman, A.M., J. Edmond, J. Seager, and G. Popják. J. Biol. Chem. 250:2045 (1975).
26. Myant, N.B., in Cholesterol Metabolism and Lipolytic Enzymes. (Ed.) J. Polonovski, Masson Publishing U.S.A., New York (1977) pp. 39-52.
27. Goldstein, J.L., and M.S. Brown. J. Biol. Chem. 249:5153 (1974).
28. Bilheimer, D.W., J.L. Goldstein, S.M. Grundy, and M.S. Brown. J. Clin. Invest. 56:1420 (1975).
29. Bhattacharyya, A.K., W.E. Connor, and A.A. Spector. J. Lab. Clin. Med. 88:202 (1976).
30. Connor, W.E., and D.S. Lin. J. Clin. Invest. 53:1062 (1974).
31. Miettinen, T.A., and A. Aro. Scand. J. Clin. Lab. Invest. 30:85 (1972).
32. Moutafis, C.D., L.A. Simons, N.B. Myant, P.W. Adams, and V. Wynn. Atherosclerosis 26:329 (1977).
33. Langer, T., W. Strober, and R.I. Levy. J. Clin Invest. 51: 1528 (1972).
34. Matthews, C.M.E. Phys. in Med. Biol. 2:36 (1957).
35. Packard, C.J., J.L.H.C. Third, J. Shepherd, A.R. Lorimer, H.G. Morgan, and T.D.V. Lawrie. Metabolism 25:995 (1976).
36. Reichl, D., L.A. Simons, and N.B. Myant. Clin. Sci. Molec. Med. 47:635 (1974).
37. Simons, LA., D. Reichl, N.B. Myant, and M. Mancini. Atherosclerosis 21:283 (1975).
38. Thompson, G.R., and N.B. Myant. Atherosclerosis 23:371 (1976).

39. Thompson, G.R., T. Spinks, A. Ranicar, and N.B. Myant. Clin. Sci. Molec. Med. 52:361 (1977).
40. Sigurdsson, G., A. Nicoll, and B. Lewis. Metabolism 26:25 (1977).
41. Thompson, G.R., A.K. Soutar, and N.B. Myant. Circulation 54, Suppl. II:26 (1976).
42. Sniderman, A.D., T.E. Carew, J.G. Chandler, and D. Steinberg. Science 183:526 (1974).
43. Levy, R.I., and T. Langer. Advanc. Exp. Med. Biol. 21:7 (1972).
44. Bachorik, P.S., J.N. Livingston, J. Cooke, and P.O. Kwiterovich. Biochem. Biophys. Res. Commun. 69:927 (1976).
45. Anderson, J.M., F.O. Nervi, and J.M. Dietschy. Biochim. Biophys. Acta 486:298 (1977).
46. Pinon, J.C., and P. Laudat. Biochim. Biophys. Acta 187:144 (1969).
47. Gotto, A.M., W.V. Brown, R.I. Levy, M.E. Birnbaumer, and D.S. Fredrickson. J. Clin. Invest. 51:1486 (1972).
48. Slack, J., and G.L. Mills. Clin. Chim. Acta 29:15 (1970).
49. Beaumont, J.-L., B. Swynghedauw, and V. Beaumont. Rev. Franc. Etud. Clin. Biol. 10:221 (1965).
50. Shattil, S.J., J.S. Bennet. R.W. Colman, and R.A. Cooper. J. Lab. Clin. Med. 89:341 (1977).
51. Howard, A.N., V. Blaton, D. Vandamme, N. Van Landschoot, and H. Peeters. Atherosclerosis 16:257 (1972).
52. Mills, G.L., C.E. Taylaur, and M.J. Chapman. Clin. Sci. Molec. Med. 51:221 (1976).
53. Jadhav, A.V., and G.R. Thompson. Clin. Sci. Molec. Med. 52:18 (1977).
54. Chatterjee, S., C.S. Sekerke, and P.O. Kwitervich. Proc. Nat. Acad. Sci. U.S. 73:4339 (1976).
55. Klauda, H.C., and D.B. Zilversmit. J. Lipid Res. 15:593. (1974).
56. Slack, J. Lancet ii:1380 (1969).
57. Heiberg, A. Clin. Genet. 9:92 (1976).
58. Buchwald, H., R.B. Moore, and R.L. Varco. Lipids 12:53 (1977).
59. Starzl, T.E., H.P. Chase, C.W. Putnam, and K.A. Porter. Lancet ii:940 (1973).
60. Starzl, T.E., H.P. Chase, C.W. Putnam, and J.J. Nora. Lancet ii:714 (1974).
61. Russell, D., C. Mieny, K.W. Heimann, M. Dinner, S.E. Levin, B.I. Joffe, D. Mendelsohn, A. Mega de Andrade, S. Lieberthal, and H.C. Seftel. Lancet ii:1205 (1976).
62. Chase, H.P., and T. Morris. Atherosclerosis 24:141 (1976)
63. Thompson, G.R., R. Lowenthal, and N.B. Myant. Lancet i:1208 (1975).
64. Barndt, R., D.H. Blankenhorn, D.W. Crawford, and S.H. Brooks. Ann. Intern. Med. 86:139 (1977).
65. Stone, N.J., R.I. Levy, D.S. Fredrickson, and J. Verter. Circulation 49:476 (1974).

BIOCHEMISTRY OF PLASMALIPOPROTEINS IN LIVER DISEASES

D.SEIDEL

Medizinische Universitätsklinik

Bergheimerstrasse 58, D-6900 HEIDELBERG, West Germany

A LIPOPROTEIN-X

The liver is the major site of synthesis of plasmalipoproteins and has central functions in lipoprotein catabolism. Since abnormal serum lipid concentrations are often associated with abnormal liver function it is reasonable to anticipate that liver disease will result in alterations of plasmalipoproteins, the protein lipid particle in which form all serum lipids circulate. Undoubtedly, the example best known to clinicians is the hypercholesterolemia and hyperphospholipidemia accompanying biliary obstruction or cholestasis, a phenomenon recorded more than a century ago by FLINT (1). The increased serum cholesterol in such patients is in form unesterified cholesterol, thus, the percentage of serum cholesterol present in esterified form is decreased, although in absolute amounts esterified cholesterol concentrations usually remain normal as long as liver function otherwise remained undisturbed, or as long as the plasma activity or the lecithin:cholesterol:acyltransferase (LCAT enzyme) is normal (2).

In recent years studies from different laboratories have well documented that the characteristic plasma lipid values found in cholestasis are preliminary due to the presence of an abnormal plasmalipoprotein, which is normally found within the LDL-density class (3, 4) and which may be isolated in intact form by combination of various physicochemical techniques (5, 6). This compound has been designed lipoprotein-X.

Lipoprotein-X (LP-X) seems to have the characteristics of a vesicle with a mean diameter of 600 Å with a tendency to aggregate as judge from negatively stained electronmicrographs at pH 7 and room temperature (7, 10).

The protein lipid composition of isolated LP-X is unique : 6% protein, 66% phospholipids, 22% unesterified cholesterol, 3% cholesterolester and 3 % triglycerides (5). The amount of bile acids bound to LP-X may very greatly causing pronounced changes in the physicochemical properties particularly in the hydrated density of the particle. The aspect has only very recently been considered. The protein portion of LP-X consists of 40% albumin, located in the core of the particle or masked by lipids and of apo-C and apo-D which are considered to be surface proteins of LP-X (6). It has been proposed that the specific combination of albumin with Apo-C peptides and Apo-D plays an important role in maintaining the structural integrity of this lipoprotein particle. In addition to the apoproteins and to albumin recent data from our laboratory demonstrate that major portions of plasma GGT are also bound to LP-X.

One important characteristic behaviour of LP-X is its electrophoretic mobility towards the cathode on agar-gel which is in contraxt to all normal plasmalipoproteins migrating to the anode. Because of its simplicity, high sensitivity and specificity to indentify LP-X, this feature has widely been used for LP-X determination and measurement.

Since the introduction of this method into clinical chemistry some years ago (12) a large body of literature uniformly documents the significance of this test in the differential diagnosis of jaundice. It separates with the highest specificity and sensitivity among all blood chemical papreters cholestasis from noncholestasis (13, 21).

It has previously been demonstrated that LP-X appears in appreciable amounts in dog and rat plasma within the first 24 hrs following surgical ligation of the common bile duct. The same time seems to be necessary for LP-X to appear in patients with estrahepatic biliary obstruction. It was also demonstrated that during extrahepatic biliary obstruction bile passes from the bile ducts to the hepatic lymphatics, from there to the thoracic lymphatic duct and finally into the blood stream. It is alo well established that liver excreeds lipids, predominantly phospholipids and unesterified cholesterol with the bile fluid, thus the lipid composition of LP-X isolated from serum shows great similarity to the lipids found in bile. Even the phosphatide distribution and phosphatide fatty acids of bile are almost identical to that of LP-X (7). These facts taken together strongly suggested a close relationship of bile lipids and LP-X.

Human bile, when subitted to agar or agarose electrophoresis reveals a lipoprotein ban migrating towards the anode, which can be visualized by polyanionprecipitation, as well as with a lipid stain or protein stain (11). Immunochemically albumin and immunoglobulins as well as other plasmaproteins can be identified but no immunoprecipitation reaction is obtained with antibodies to any of the major apolipoproteins Apo-A, Apo-B, Apo-C, apo-D and apo-E. Under the electronmicroscope it is not possible to detect reproducable any particular structures occasionally some lamellar structures are apparent. LP-X, however, with its characteristic structure under the electronmicroscope or even more specifically with its typical migration towards the cathode on agar electrophoresis cannot be detected in neither native nor water dialyzed bile. All bile lipids may be separated from bile in form of an albumin-lipid complex (designated bile LP). This complex shows a lipid composition very similar to LP-X but displays different physico-chemical characteristics. When bile, however, is incubated in vitro with either pure albumin or with total (LP-X negative) serum, LP-X is formed and may be isolated showing an electrophoretic behaviour and other physochemical characteristics similar to that found in the serum of patients suffering from cholestatic liver disease (11). Addition of various amounts of isolated VLDL, LDL or HDL as well as purified immunoglobulins to native bile does not result in LP-X formation.

The amount of LP-X formed from bile after addition of albumin or serum depends to a high degree on the concentration of native bile, in particular on the concentration of bile lipoprotein, bile salts and the amount of albumin added tot he system. Comparison of LP-X formed by albumin alone (designated LP-X alb) or by serum (LP-X serum) with native LP-X revealed a very similar lipid composition. Significant differences are apparent, however, in their protein moiety and the structure properties as judged by electronmicroscopy and electrophoresis on agar-gel. LP-X alb shows somewhat larger and more heterogenous particles on electronmicrographs, it migrates further to the cathode on agar electrophoresis as does LP-X and is devoid of Apo-C and Apo-D. After in vitro incubation with HDL or VLDL this material as well as LP-X serum shows these apolipoproteins, the latter is almost indistinguishable from native LP-X. From these in vitro experiments which are strongly supported by in vivo studies in dog it seems evident that LP-X formation takes place when bile LP is converted into LP-X by the action of albumin in the proper ratio to bile salts. Thus albumin, rather than other Apo-X peptides seems to be the key protein in LP-X formation and in maintaining the unique physico-chemical characteristics of LP-X, although Apo-C and Apo-D in addition may have some stabilizing effect on the structure of this abnormal plasma lipoprotein.

The plasma influx of biliary phospholipids and free cholesterol in form of bile lipoprotein or LP-X respectively may be of help for our understanding of cholesterol metabolism and the mechanism of hypercholesterolemia in cholestasis. It is well established that cholesterol synthesis but not so phospholipids synthesis is increased in cholestasis. Moreover, phospholipids in a high ratio to cholesterol as in LP-X stimulate hepatic cholesterol synthesis and recent results from our laboratory show that LP-X, probably because it is devoid of Apo-B, is not able to inhibit hepatic cholesterol synthesis in contrast to Apo-B carrying lipoproteins (22).

The role of lecithin:cholesterol:acyltransferase, the LCAT-enzyme in the metabolism of plasmalipoproteins is still a matter of speculation. This holds particularly for the cholestatic conditions. LCAT is likely to be produced in the liver and the lipoprotein alterations in liver disease are in many respects similar to those found in familial LCAT-deficiency. Although some experimental data are available (23) on the relationship of LCAT and lipoproteins in liver disease, our knowledge in this field is still very lipited. The primary reason for this seems to be that the exact mode of action of this enzyme, which is responsible for esterification of lipoprotein cholesterol in plasma is still not clear. It was believed that HDL or lipids in HDL are the major cholesterol and lecithin substrate for the enzyme and that Apo-A_1 was the primary cofactor. However, there are also reports indicating that VLDL rather than HDL should be considered as substrate and that Apo-D or even a protein in the d 1.25 g/ml fraction are the most potent activators for the enzyme. It is without doubt that cholesterol, lecithin and a protein activator is required for the LCAT reaction but more than that the structural arrangements of these moieties within the lipoprotein particle must carefully be considered. Evidence for the great importance of this aspect is provided by studies of interactions between membrane lipids and lipolytic enzymes. The relationship, if there is any, between LP-X and LCAT in cholestasis is not clear. LCAT-activity may be normal elevated or decreased in cholestasis despite the presence of LP-X. According to the carefully performed studies by RITLABD and GJONE (24) an inverse correlation between LCAT-activity holds only for the concentration of plasma cholesterolester concentration but not for LP-X concentration in neither cholestatic liver disease nor in familial LCAT deficiency. The fact that LP-X is often the only plasmalipoprotein abnormality particularly in forms of mild cholestasis accompanied with an elevated LCAT-activity and normal ester ratios in the other lipoproteins may indicate that LP-X is not a particularly good substrate for the LCAT enzyme. Our own in vitro studies in which isolated LP-X was incubated with a partially purified LCAT fraction supports this idea and are in agreement with data by MAGNANI (25). It may be speculated that the mechanism of LP-X formation in familial type of LCAT deficiency is entirely different from that in cholestasis. It may, however, also be a result of the particular physicochemical conditions of the lipids in this disease.

B HYPERTRIGLYCERIDEMIA IN LIVER DISEASE

Hypertriglyceridemia as a consequence of alcohol intake is a well known phenomenon (26) which, however, is not fully understood in the pathophysiological mechanisms.

Only recently particular attention has been paid to the hypertriglyceridemia, which may be found in some patients with liver dysfunction (27, 31), a metabolic situation most frequently accompanied by severe cholestasis. It was demonstrated that the majority of triglycerides of such patients is found within the LDL lipoprotein fraction. The isolation of a large (300-700 Å in diameter) triglyceride rich low density lipoprotein (d 1.019 - 1.063 g/ml) different from LP-X and normal beta-lipoproteins was later achieved by a combination of ultracentrifugation, cold ethanol fractionation and affinity chromatography (31). This lipoprotein has been designated $beta_2$-lipoprotein because it shares certain physicochemical characteristics and chemical properties with normal beta-lipoproteins. Detailed analyses of the lipoprotein families of such patients indicated that the increased concentration of LDL triglycerides as well as the decreased VLDL-TG/ 1.006 g/ml-TG ratio and hereby the hypertriglyceridemia of such patients with liver disease is primarily due to the presence of $beta_2$-LP. Since the protein moiety of this low density lipoprotein consists of Apo-B and Apo-C and possibly also of trace amounts of Apo-A it is likely that $beta_2$-LP derives from lipoproteins of lower density which are known to contain all major apoproteins. This possibility is supported by a dietary experiment (31). The decrease of its plasma concentration on a low fat diet may indicate its nature as a degradation product or intermediate particle of chylomicron metabolism, although an additional relationship to VLDL catabolism cannot be ruled out with our present knowledge. One significant mechanism responsible for the accumulation of this low density lipoprotein is the markedly reduced hepatic lipase activity, which we first demonstrated in such patients (31) and which may be relevant for the mechanism of their hypertriglyceridemia.

C CHANGES OF THE VLDL AND HDL FRACTION OCCURING IN LIVER DISEASE

Besides striking abnormalities of the LDL-fraction changes from normal can also be found in other density classes (32). Using the analytical ultracentrifuge a diminuition of the HDL-fraction was demonstrated. The lipoprotein electrophoresis is frequently altered showing only one broad beta-band in beta-position, the pre-beta- and alpha-bands are lacking (32).

The absence of pre-beta- and alpha-lipoprotein bands on lipoprotein electropherograms does, however, not necessarily allow the conclusion that the corresponding lipoprotein density fractions are absent in the plasma sample.

The VLDL-fraction may even be increased in such patients. Isolated VLDL exhibit not pre-beta- but beta-mobility on agarose-electrophoresis and show a marked decrease of Apo-C in the apoprotein moiety. Incubation of VLDL isoloated from liver diseased patients with normal HDL restores the normal pre-beta-mobility, indicating a transfer most likely of apoproteins from HDL to VLDL.

The lacking alpha-lipoprotein band is usually accompanied by a decrease of the HDL-concentration to about 75% on the average. The plasma concentration of A-I and A-II together, however, is not found to be decreased that much. The immunoelectrophoretical pattern of HDL from liver diseased patients against alpha-lipoproteins strongly suggests a dissociation of apo-A-I and apo-A-II forming two non identical protein molecules and corresponding lipoprotein families. Such HDL fractions hardly contain any cholesterol or triglycerides but phospholipids.

In addition to the altered alpha-lipoproteins a second abnormal lipoprotein has been isolated from the HDL-fraction of patients suffering from severe prolonged cholestasis. It was designated LP-E, showing primarly if not only the arginine-rich apoprotein (34). On electonmicrographs it exhibits rouleaux formations like LP-X and migrates in beta-position on agarose electrophoresis. In contrast to LP-X it migrates also on agar towards the anode.

Since A-I and A-II are impaired to bind neutral lipids in dissociated form they are detectable on lipoprotein electrophoresis. The lack of neutral lipids may also cause the deficiency of HDL in LP-C and may, in addition, be of importance with regard to the regulation of chylomicron and VLDL catabolism. To clarify the nature of the structural abnormality of high density lipoproteins in liver disease much further work needs to be done. The concentration of A-I and A-II has to be determined in the plasma of these patients and under the course of the disease. Data of such studies should be related to other parameters of lipid and lipoprotein metabolism.

REFERENCES

1. FLINT, A. jr. : Experimental researches into a new excretory function of the liver, consisting in the removal of cholesterine from the blood, and its discharge from the body in the form of stercorine.
   Amer. J. med. Sci. 44, 305 (1962).
2. SEIDEL, D. : Hyperlipoproteinaemia bei Erkrankungen der Leber. Schweiz. med. Wschr. 105, 857 (1975).
3. RUSS, E.M., RAYMUNT, J. and D.P.BARR : Lipoproteins in primary biliary cirrhosis.
   J. Clin. Invest. 35, 133 (1956).
4. SWITZER, S. : Plasma lipoproteins in liver disease. I. Immunologically distinct low-density lipoproteins in patients with biliary obstruction.
   J. Clin. Invest. 46, 1855 (1967).
5. SEIDEL, D. ALAUPOVIC, P. and R.H. FURMAN : A lipoprotein characterizing obstructive jaundice. I. Method for quantitative separation and identification of lipoproteins in jaundiced subjects. J. Clin. Invest. 48, 1211 (1969).

6. SEIDEL, D. ALAUPOVIC, P., R.H. FURMAN and W.J. McCONATHY : A lipoprotein characterizing obstructive jaundice. II. Isolation and partial characterization of the protein moieties of low density lipoproteins.
J. Clin. Invest. 49, 2396 (1970).

7. PICARD, J. and D. VEISSIERE : Séparation des lipoprotéines sériques anormales dans la cholestase.
C.R. Acad. Sci. (Paris) Ser. D. 270, 1845 (1970).

8. MILLS, G.L., SEIDEL, D. and P.ALAUPOVIC : Ultracentrifugal characterization of a lipoprotein occuring in obstructive jaundice.
Clin. Chim. Acta 26, 239 (1969).

9. HAMILTON, R.L., HAVEL, R.J. KANE, J.P., BLAUROCK, E.A. and T.SATA : Cholestasis : lamellar structure of the abnormal human serum lipoprotein.
Science (Wash. D.C.) 172, 475 (1971).

10. SEIDEL, D., AGOSTINI, B. and P.MUELLER : Structure of an abnormal plasma lipoprotein (LP-X) characterizing obstructive jaundice.
Biochim. Biophys. Acta 260, 146 (1972).

11. NANZATO, E., FELLIN, R., BAGGIO, G., WALCH, S., NEUBECK, W. and D. SEIDEL : Formation of lipoprotein-X. Its relationship to bile compounds.
J. Clin. Invest. 57, 1248 (1976).

12. NEUBECK, W. and D. SEIDEL : Direct method for measuring lipoprotein-X in serum.
Clin. Chem. 21, 853 (1975).

13. SEIDEL, D., SCHMITT, E.A. and P. ALAUPOVIC : Ein abnormes Low-density-Lipoprotein bei Cholestase. II. Bedeutung in der Differentialdiagnose des Ikterus.
Dtsch. Med. Wochenschr. 95, 1805 (1970).

14. POLEY, J.R., ALAUPOVIC, P., McCONATHY, W.J., SEIDEL, D., ROY, C.C. and A. WEBER : Diagnosis of extrahepatic biliary obstruction in infants by immunochemical detection of LP-X and modified 133 I-Rose Bengal excretion.
J. Lab. Clin. Med. 81, 325 (1973).

15. SEIDEL, D., GRETZ, H. and C. RUPPERT : Significance of the LP-X test in differential diagnosis of jaundice.
Clin. Chem. 19, 86 (1973).

16. PREXL, H-J. and W. PETEK : Die Bedeutung des Lipoprotein-X und der Serumcholestase in der präoperativen Diagnostik des Verschlussikterus.
Chirurg. 44, 310 (1973).

17. RITLAND, S., BLOMHOFF, J.P., ELGJO, K. and E. GJONE : Lipoprotein-X (LP-X) in liver disease.

18. VERGANI, C., PIETROGRANDE, M., GRONDONA, M.C. and M. PIZZOLATO : Study of the abnormal lipoprotein-X in obstructive and non-obstructive jaundice.
Clin. Chim. Acta 48, 243 (1973).

19. ROSTI, D. and C.G. DE GASPARI : Valoré diagnostico della LP-X negli itteri neonatali.
Minerva Pediatr. 26, 582 (1974).

20. MAYER, K. : Der Wert der LP-X-Bestimmung zur Erfassung einer Cholestase.
In : Jahrestagung der Deutschen, Österreichischen und Schweizerichen Gesellschaft für klinische Chemie, Freiburg, West Germany, März 16-13 (Abstr.) 1975).

21. FISCHER, M., FALKENSAMMER, C., BARONACH, G., WAKETUCH, S., KRONBERGER, O. and H. SCHNACK : Zur Diagnose der Cholestase : Lipoprotein-X (LP-X).
Wien. Klin. Wochenschrift 87, 524 (1975).

22. LIERSCH, M., HEUCK, C.C., BAGGIO, G. and D. SEIDEL : Regulation of Cholesterol Biosynthesis in Rat Liver by Two Different Plasma Lipoproteins.
Atherosclerosis 26, 505 (1977).

23. McINTYRE, N., HARRY, D.S. and A.I.G. PEARSON : The Hypercholesterolaemia of Obstructive Jaundice.
Gut 16, 379 (1975).

24. RITLAND, S., BLOMHOFF, J.P. and E. GJONE : Lecithin:Cholesterol Acyltransferase and Lipoprotein-X in Liver Disease.
Clin. Chim. Acta 49, 251 (1973).

25. MAGNANI, H.N. : The influence of LP-X and other lipoproteins associated with hepatic dysfunction on the activity of lecithin: cholesterol acyltransferase.
Biochem. Biophys. Acta 450, 390 (1976).

26. BARAONA, E. and LIEBER, L.S. : Effects of chronic ethanol feeding on serum lipoprotein metabolism in the rat.
J. Clin. Invest. 49, 769 (1970).

27. KLÖR, U., DITSCHUNEIT, H.H., RABOW, D. and H.DITSCHNUNEIT : Further characterization of dyslipoproteinemia in hepatic disease.
Abstr. Eur. J. Clin. Invest. 2, 291 (1972).

28. ALCINDOR, L.G., INFANTE, R. and J.CAROLI : Plasma VLDL catabolism in cholestasis. Abstr. 5th meeting of the int. Ass. for the study of the liver. Versailles, July 1972.

29. PEARSON, A.J.G. : Triglycerides in obstructive liver disease.
Abstr. 5th meeting of the int. Ass. for the study of the liver.
Versailles, July 1972.

30. FELLIN, R. and D. SEIDEL : Behaviour of serum lipoproteins in cholestasis.

1. Int. Symp. on cholestasis, Florence, June 1973.
Proc. of the Symposium.

31. MÜLLER, P., FELLIN, R., LAMBRECHT, J., AGOSTINI, B., WIELAND, H., ROST, W. and D. SEIDEL : Hypertriglyceridemia Secondary to Liver Disease.
Europ. J. Clin. Invest. 4, 419 (1974).

32. SEIDEL, D. GRETEN, H., GEISEN, H.P., WENGELER, H. and H.WIELAND : Further aspects in the characterization of high and very low density lipoproteins in patients with liver disease.
Europ. J. Clin. Invest. 2, 359 (1972).

33. GOFMAN, J. : The serum lipoprotein transport system in health, metabolic disorders, atherosclerosis and coronary artery diseases.
Plasma 2, 484 (1954).

34. SEIDEL, D. et al. : unpublished observations.

THE ROLE OF LIPOPROTEINS IN HUMAN ATHEROSCLEROSIS (Lecture I)

K.W. Walton

Department of Experimental Pathology
University of Birmingham
Birmingham 15 England

The atherosclerotic plaque at all stages of its development, and wherever it occurs in the arterial tree, is characterised by lipid deposition. Earlier suggestions that the lipid is derived solely from breakdown ("fatty metamorphosis") of intrinsic components of the arterial wall (1), or originates from blood clots or cellular components of the blood deposited on the endothelial surface and later incorporated into the wall (2), are no longer widely entertained (for review, see Ref.3). Instead, current research seems broadly to support the hypothesis of Virchow (4) that the lipid in atherosclerotic lesions derives from the "insudation" of plasma into the arterial wall. However the original hypothesis requires modification and restatement in terms of modern knowledge to take account of (a) what is now known of the forms in which lipid is transported in the blood; (b) the known association of risk factors with the occurrence of atherosclerosis; and (c) the mechanisms whereby some of the more important risk factors operate at the level of the arterial wall. The present communication will deal with the role of lipoproteins in the evolution of the atherosclerotic plaque. The accompanying communications will deal with the influence of other factors on this process.

Evidence implicating lipoproteins in atherogenesis

Indirect (correlative) evidence relating the serum levels of low-density (LDL), intermediate (IDL) and very-low density lipoproteins (VLDL) - hereafter collectively referred to as total low-density lipoproteins, or TLDL - with the occurrence of atherosclerosis with respect to age, sex and geographical location in health, and also with respect to diseases accompanied by hyperlipidaemia, has been reviewed elsewhere (3, 5-7).

Direct evidence of the implication of TLDL in atherosclerotic lesions has been obtained by (i) the use of istopically-labelled lipoproteins; (ii) immuno-histochemical techniques (iii) examination of arterial extracts and (iv) by a combination of (ii) and (iii).

(i) Use of radioisotopically labelled lipoprotein. In certain instances in which autologous TLDL, radioactively labelled in its protein moiety (apolipoprotein) has been administered in vivo for turnover studies, and subsequent autopsy examination has been possible, it has been shown (8, 9) that atherosclerotic vessels accumulate radioactivity at a higher level per gram of tissue than other organs, suggesting direct entry of the labelled lipoprotein into the affected sites.

(ii) Immunohistochemical evidence. Using the technique of immunofluorescence and antisera to: intact LDL (subsequently shown to react mainly or exclusively with apolipoprotein B, an antigenic component common to LDL, I.D.L. and VLDL); and to intact high-density lipoprotein (HDL), the presence of TLDL but not HDL in atherosclerotic lesions was inferred by several authors (for references, see Ref.31. It was observed that the distribution of specific fluorescence for apolipoprotein B in lesions corresponded exactly with that of extracellular lipid (as delineated by conventional fat stains) when the same section was examined sequentially by both techniques (10). Antibodies to lipoproteins are directed mainly or exclusively against the protein part of the molecule, whereas fat "stains" such as Oil red O and the Sudan dyes are selectively soluble in lipids but unreactive with proteins. Where precise identity of distribution between lipid "staining" and immunochemical reaction for TLDL was found, therefore, it appeared reasonable to infer that the intact lipoprotein molecule must be present.

On the other hand, in relation to fat-filled cells within lesions, it was found that some such cells gave bright specific fluorescence for apolipoprotein B, others reacted weakly and some failed to react. On the same basis as above, it was inferred that full immunoreactivity was compatible with the uptake of intact lipoprotein by phagocytic cells whereas relative lack, or loss, of immunoreactivity was suggestive of progressive degradation of the lipoprotein molecule by intracellular proteases. These, it was considered, first digest the apolipoprotein (site of antigenicity) leaving a lipid residue in the cell (also possibly contributed to by local lipid synthesis of triglyceride and phosopholipid) detectable by conventional fat "stains". The appearance of fat-filled cells was thus thought to indicate the activity of a cellular removal mechanism.

Using antisera specific for other individual plasma proteins the only one found (11) significantly to be related to atherosclerotic lesions was fibrinogen, in confirmation of other observations (12-14).

In later work, antisera to additional antigenic components of lipoproteins have been employed. These have included antisera to apolipoprotein C and to the arginine-rich peptide (components present in IDL and VLDL) and to the peptide characteristic of Lp(a) lipoprotein, a variant lipoprotein easily detectable in the serum of some persons but not others (15-17). Identity of distribution in plaques was found between that of apolipoproteins B and C or ARP, and also between apolipoproteins B and Lp(a), suggesting that lipoproteins even larger in molecular size than LDL itself participate in atherogenesis (also see below).

This suggestion was supported by immunohistochemical examination of plaques using antiserum to apolipoprotein B labelled with horse-radish peroxidase instead of, or in parallel with, fluorescein-labelled antisera (18, 19). Antibodies labelled with horse-radish peroxidase and bound to TLDL in tissues, when treated with diaminobenzidine, give rise to a brown reaction product allowing the localisation of the lipoprotein in tissue sections in the light microscope. Following treatment with osmic acid, an electron-dense reaction product is obtained allowing localisation also in the electron microscope. On using this technique on plaques from the aorta, basilar, coeliac axis and splenic arteries (19) areas containing pools of amorphous lipid (reactive also with fluorescein-labelled anti-apolipoprotein B) in the interstitial matrix were examined. In the electron microscope large numbers of spherical particles coated by electron dense reaction product were seen. In some instances, the plane of section was such as to show granules of reaction product arranged in circles around a core of greyish lipid material, as though surrounding the circumference in section of a sphere. The internal diameters of these structures was found to vary between about 22 and 48 nm. This corresponds to the range of molecular diameters found for individual lipoprotein molecules when isolated and purified fractions of LDL, IDL and VLDL are examined in the electron microscope. (20) It would seem therefore, from direct visualisation, that molecules within the whole size range encompassed by TLDL can participate in lesions.

(iii) Examination of arterial extracts. Extracts of the arterial wall prepared by homogenization, sonication, or by the use of a tissue-press have allowed the identification of lipoproteins in the tissue fluids by immunoprecipitation (20-22). Using a technique of electrophoresing the proteins in the interstitial fluid of arterial intima directly into antiserum-containing gels, it has been demonstrated (23-25) that the amount of TLDL recoverable from the intima is related to the level of serum lipids and lipoproteins in life. It has also been shown that fibrinogen is similarly recoverable from intima and that a proportion of this is clottable with thrombin (25).

(iv) Combined techniques. On applying the method of two-dimensional electrophoresis to remove and characterise the proteins present in the arterial intima, it could be demonstrated that almost all the proteins present in plasma are present in the interstitial fluid. But if intima from atherosclerotic lesions is thus extracted until no further protein emerges and the same intimal sample is then examined by conventional histology and by immunofluorescence, then a "firmly bound" fraction of lipid and lipoprotein is still demonstrable in the tissues (26).

On applying some of the above-mentioned techniques to areas unaffected by visible plaque formation, the results obtained vary with the age of the subject in a manner suggesting variability of permeability of the arterial wall, with age, to plasma proteins of differing molecular size (3). Other factors causing selective or diffuse alteration of vascular permeability under various pathological circumstances are discussed in Lecture II.

The evidence discussed so far may be summarised as indicating that the development of atherosclerotic lesions depends not only upon the entry of plasma proteins into areas of increased permeability, but also upon the selective binding at such sites of these lipoproteins (and fibrinogen) to components of the arterial wall.

Mechanism, sites of binding, and effect of entrapment of TLDL in the arterial wall

The mechanisms variously proposed for retention of the 'firmly bound' fraction of TLDL in the arterial wall have been previously reviewed and discussed (3, 27). Evidence for the view gaining widest acceptance, namely that TLDL and fibrinogen are retained in the arterial wall by their interaction with the sulphated glycosaminoglycans (S-GAG's) or proteoglycans (S-PG's) of the connective tissues of the arterial wall, can be summarised as follows :

(i) In relation to lipoproteins, selectivity of retention of TLDL but not HDL implies that it is the nature of the vehicle for lipids (some aspect of the physicochemical characteristics of TLDL) rather than the nature of the lipids being carried, which determines initial localisation. Since fibrinogen is also present in lesions, one might suspect both TLDL and fibrinogen to share some common characteristic of reactivity. In this context it seems of significance that among the plasma proteins, only TLDL and fibrinogen selectively form insoluble complexes (co-acervates) on the addition of charged polysaccharides, S-GAG's, or S-PG's derived from the arterial wall, under *in vitro* conditions (for references, see Refs 3 and 27).

(ii) In the intact artery, histochemical methods and radiosulphate uptake show a significant correlation between the distribution of S-GAG's or S-PG's and of LDL and fibrinogen in plaques (28, 29).

(iii) Even "mucoid elevations" of the intima - very early lesions which some authors (3o, 31) hold to precede visible lipid (lipoprotein) accumulation in the wall - show local accumulation of S-GAG's. Moreover, many of the stimuli giving rise to increased vascular permeability also produce a "mesenchymal reaction" (32) in the subendothelial tissues of arteries, associated with increased radiosulphate uptake, suggesting increased production of S-GAG's. If the conditions giving rise to altered permeability also promote synthesis of the components which are reactive with and precipitate TLDL and fibrinogen,, it is possible to see how and why selective entrapment of these proteins is initiated.

The S-GAG's and S-PG's of connective tissues are intimately involved in the structural organisation of collagen and elastin. During fibrillogenesis they provide a framework allowing the orientation and steric arrangement of developing micro-fibrils to allow their cross-linking and aggregation. Fine structural studies (18, 19) using the immunoperoxidase technique have shown that TLDL molecules can be identified on the outer surfaces of bundles of micro-fibrils forming the collagen fibre seen in the light microscope and thus conforming with the "peri-fibrous" distribution (33) seen by conventional microscopy. But TLDL binding to individual fibrils of native (mature) collagen and of elastin was also demonstrable.

In advanced plaques, TLDL binding was also found in relation to an 'abnormal' polymorphic form of collagen, known as fibrous long-spacing collagen (FLSC). In this instance, the TLDL was bound at the prominent transverse bands occurring at intervals of about 120 nm along these along these structures. It is precisely at these sites where interfibrillar bridging by S-GAG's has been shown to occur (34).

Significance of binding of TLDL

The entrainment of TLDL in the intimal gel often evokes a local cellular reaction. This may serve, as previously suggested, as a removal mechanism. But alternatively, the degradation of the lipoprotein and release from it of free and esterified cholesterol, phospholipids and other components may stimulate further cellular reaction and fibrosis. Alternatively a mixture of intact and degraded lipoprotein, probably contributed to by local triglyceride and phospholipid synthesis, may result in the accumulation of "atheromatous gruel" in the plaque, making it liable to ulceration, superadded thrombosis and the later complications of atherosclerosis.

The finding of TLDL binding to FLSC may also be of significance since it has been reported (35) that when arterial explants are maintained in tissue culture, the formation of this 'abnormal' form of collagen occurs in the presence of hyperlipidaemic but not normolipidaemic serum in the medium. The tensile strength of FLSC is likely to be less than that of 'normal' collagen so that its occurrence may be conducive to arterial dilatation, aneurysm formation and other complications associated with reduced tensile strength of the arterial wall (19)

REFERENCES

1. Thoma, R. (1883) Virchows Arch. Pathol. Anat. Physiol., 93, 443.

2. Rokitansky, C.von (1952) In A Manual of Pathological Anatomy (translated by G.E. Day), Vol 4, p.261. Sydenham Society, London.

3. Walton, K.W. (1975) Amer. J. Cardiol., 35, 542.

4. Virchow, R. (1862) In Gesammelte Abhandlungen zur wissenschaftlichen Medizin, Max Hirsch, Berlin.

5. Walton, K.W. (1969) In The Biological Basis of Medicine (Eds., Bittar, E. and Bittar, N.) Vol.6. pp 193-223, Academic Press, New York.

6. Walton, K.W. (1973) In Textbook of Geriatric Medicine and Gerontology, (Ed. Brocklehurst, J.C.) pp.77-112, Churchill Livingstone, Edinburgh.

7. Walton, K.W. (1974) Adv. Drug Research, 9, 55.

8. Walton, K.W., Scott, P.J., Verrier Jones, J., Fletcher, R.F. and Whitehead, T.P. (1963) J. Atheroscler. Res., 3, 396.

9. Scott, P.J. and Hurley, P.J. (1970) Atherosclerosis, 11, 77.

10. Walton, K.W., Williamson, N. and Johnson, A.G., (1970) J. Pathol., 101, 205.

11. Walton, K.W. and Williamson, N. (1968) J. Atheroscler., 8, 599.

12. Woolf, N. and Crawford, T. (1960) J. Pathol. Bact., 80, 405.

13. Haust, M.D., Wyllie, J.C. and More R.H. (1964) Amer. J. Pathol, 44, 255.

14. Wyllie, J.C., More, R.H. and Haust, M.D., (1964) J. Pathol. Bact., 88, 335.

15. Walton, K.W. In Atherosclerosis III, (Eds. Schettler, G. and Weizel, A), pp. 93-95, Springer Verlag, Berlin.

16. Walton, K.W., Morris, C.J. and Pagnan, A. - in preparation.

17. Walton, K.W., Hitchens, J., Magnani, H.N. and Khan, M.K. (1974) Atherosclerosis, 20, 323.

18. Hoff, H.F. and Gaubatz, J.W. (1975) Virchows Arch. Pathol. Anat. Histol., 369, 111.

19. Walton, K.W. and Morris, C.J. In Proceedings of First International Atherosclerosis Conference, Austrian Association for Morphological and Functional Research in Atherosclerosis, Vienna, 1977, (ed. H. Sinzinger) Karger, Basel, in press.

20. Ott, H., Lohss, F. and Gergely, J., (1958) Klin. Wschr., 36,383

21. Tracey, R.E., Merchant, E.B. and Kao, V.C. (1961) Circ.Res., 9, 472.

22. Klimov A.N., Denishenko, A.D. and Magracheva, E.Y. (1974) Atherosclerosis, 19, 243.

23. Smith, E.B. and Slater, R.S. (1971) Biochem. J., 123, 39P.

24. Smith, E.B. and Slater, R.S. (1972) Lancet, 1, 463.

25 Smith, E.B., Slater R.S. and Hunter, J.A. (1973) Atherosclerosis, 18, 479

26. Walton, K.W. and Bradby, G.V.H. In Proceedings of International Workshop-Conference on Atherosclerosis, London, Ontario, 1975, (ed. M.D. Haust), Plenum Press, New York, in press.

27. Walton, K.W. (1974) In Atherosclerosis III. (Eds. Schettler,G. and Weizel, A.) pp. 215-217, Springer Verlag, Berlin.

28. Curran, R.C. and Crane, W.A.J. (1962) J. Pathol. Bact. 84,405.

29. Gerő,S., Virag, S., Bihari-Varga, M., Szekely. J. and Feher, J. (1967) In Progress in Biochemical Pharmacology (Eds., Kritchevsky, D., Paoletti, R. and Steinberg, D.) Vol. 2, p.290, Karger, Basel.

30. Haust, M.D. (1971) Human Pathol., 2, 1.

31. Smith, E.B. and Slater, R.S. (1973). In Atherogenesis : Initiating Factors. Ciba Foundation Symposia, New Series, No.12, p.39, Excerpta Medica, Amsterdam.

32. Hauss, W.H., Junge-Hülsing, G. and Hollander, H.J. (1962) J. Atheroscler. Res., 2, 50.

33. Smith, E.B., Evans, P.H. and Downham, M.D. (1967) J. Atheroscler. Res., 7, 171.

34. Doyle, B.B., Hukins, D.W.L., Hulmes, D.J.S., Miller, A. and Woodhead-Galloway, J. (1975) J. Molec. Biol., 79-99.

35. Larrue, J. Darel, D., Demond. J. and Bricaud, H. Atherosclerosis - in press.

THE INTERACTION BETWEEN HYPERLIPIDAEMIA AND OTHER RISK FACTORS IN THE PATHOGENESIS OF ATHEROSCLEROSIS (LECTURE II)

K.W. Walton

Department of Experimental Pathology
University of Birmingham
Birmingham 15 England

In my first lecture evidence derived from application of a variety of techniques was reviewed which suggests that atherosclerotic lesions arise in general because certain macromolecular plasma proteins (most notably TLDL and fibrinogen) interact with components of the connective tissues of the arterial wall. In the present lecture, evidence will be considered which points to this being a general mechanism which even affects extravascular connective tissues at certain sites.

If a modified insudative mechanism of atherogenesis is provisionally accepted, then the composition of the plasma will clearly be of importance. But it is often objected that the composition of the blood in respect of TLDL and fibrinogen concentrations is uniform throughout the vascular tree and yet lesions occur discontinuously and only at certain sites of predilection in arteries and not in reins. It is also objected that atherosclerosis may be encountered in individuals showing minimal or no elevation of serum lipid levels (1).

In seeking to answer these objections it is necessary:
(a) to distinguish between the factors determining the localisation and those increasing the severity of lesions;
(b) to appreciate how certain factors mutually re-inforce one another in a fashion which is not simply additive;
(c) to seek understanding of the underlying mechanisms by examining sites other than the arterial plaque itself.

It is suggested that the prime factor determining the localisation of lesions is alteration of permeability of the vessel wall at the sites affected.

General considerations relating to variation of vascular permeability

The relative permeability of endothelium by different plasma proteins, and therefore the distribution of given protein between intravascular and extravascular compartments, is probably largely determined by the molecular size and shape of the protein. For example, in the case of a relatively small molecule like serum albumin, isotopic turnover studies have established that 1½ - 2 times as much albumin is distributed extravacularly as is present intravascularly (2,3). On the other hand, in the case of TLDL, similar methods have shown that, of the total body pool, about two-thirds is intravascular and only one-third (much of which is probably accounted for by the liver pool) is extravascular (4-8). Moreover, labelled TLDL molecules introduced into the circulation have been found to reach full equilibrium very slowly with the extravascular pool (8).

These observations lead to the conclusion that TLDL is normally largely confined to the circulation and that demonstration of its presence (and in particular of VLDL, the largest of the size-range of these molecules) in the arterial wall (a site outside the circulation) can be logically regarded as an indication that endothelial permeability is selectively altered at such a site.

Selective alteration of permeability

In experimental animals naturally prone to the development of atherosclerosis, such as the pig, it has been shown repeatedly that areas, in which lesions develop preferentially, at an earlier stage show selective alteration of permeability in allowing the ingress to the same areas of a variety of marker substances such as dye-labelled proteins, colloidal iron, colloidal lipid and even finely particulate material such as ferritin or horse-radish peroxidase (for references, see Ref. 9 and 1¼).

With regard to the mechanism of production of localised alteration of permeability at such sites of predilection, it seems reasonable to assume that the arterial wall, like other living connective tissues, can react to stimulation by the local release of histamine, kinins, and other intrinsic mediators of altered permeability. Whether the 'trauma' of haemodynamic stresses at bends, flow-dividers etc. in the arterial tree can operate in this way is difficult to determine in the human. Experimentally, local trauma to elective sites, applied either internally or externally to the artery (for reviews, see 11,12) causes initial oedema. Later, lipid infiltration has been observed even in normolipidaemic animals but is accelerated if hyperlipoproteinaemia is induced (13).

Localised alterations which are widespread in the arterial tree may be immunologically mediated by allergic injury to arteries.

For example, the induction of immune complex disease in lipid-fed rabbits increases the extent and severity of lipid-filled arterial injuries (14). However, it has been noted that in such experiments lesions can be observed at sites not usually involved in animals subjected to lipid-feeding alone (15). This is probably because immune complexes have their own characteristic sites of lodgement (in small arteries) which are not identical with the common sites of atherosclerotic involvement (in large arteries). While it can be envisaged that immune complexes (and especially those incorporating activated complement and involving blood platelets) may damage endothelium, in personal experiments (Valente and Walton, unpublished) it has been found difficult to demonstrate the simultaneous presence of components of such complexes, and of lipoprotein infiltration in lesions in this experimental model. This is probably because the rate of removal of the components of immune complexes is rapid in relation to the time-scale of production of unequivocal atherosclerotic lesions. However, it is possible that a human counterpart to the experimental situation is seen in the increased speed of development and severity of atherosclerosis seen in the vessels of transplanted organs undergoing rejection and also showing evidence of immune arteritis.

Widespread but segmental alteration of permeability can be envisaged as occurring in arterial territories involved by inflammatory processes. This may account for the observed tendency for atherosclerosis to be more severe when it supervenes in arterial segments affected by various forms of granulomatous arteritis such as temporal arteritis, Takayasu's disease or syphilitic arteritis (16 - 19). Granulomatous arterial disease also affects the outflow from the wall because of obstruction of lymphatic drainage (9).

Diffuse alteration of vascular permeability

In the previous lecture, attention was drawn to the generalised increase in vascular permeability which accompanies ageing. This is given its fullest expression in the 8th and 9th decades of life since the thickened and pigmented aortic intima of individuals in these age groups, even in areas apparently unaffected by discrete atherosclerotic lesions, shows diffuse sub-intimal infiltration by lipid in the form of TLDL (20).

In acute experiments, the intravenous injection of histamine epinephrine and other vaso-active substances produces a generalised "oedematous intimal reaction' attended later by the insudation of TLDL into the intima (21).

Hypoxia was suggested by Hueper (22, 23) as a factor inducing diffuse alteration of vascular permeability so as to allow the influx into the arterial wall of both exogenous and endogenous molecules (including the plasma lipids) from the circulation. Experimentally, the maintenance of lipid-fed hyperlipoproteinaemic animals at reduced oxygen tensions (24) or in atmospheres containing significant concentration of carbon monoxide (25) accelerates and intensifies experimental atherosclerosis.

The human counterpart of this is seen in the heavy cigarette smoker (26) who may show a consistent level of 10-15 per cent of carboxyhaemglobin in the blood (i.e., a level found to enhance atherosclerosis in the experimental animal). It seems likely that this is one of the underlying mechanisms leading to the recognition of heavy smoking as a risk factor (27). It has been observed that diffuse sub-intimal lipid (TLDL) infiltration in areas unaffected by visible raised lesions, and resembling that otherwise seen in the last decades, is observed precociously in the arteries of young males who were heavy smokers in life (20).

At the ultrastructural level, in animals subjected to hypoxia, the injection of vaso-active compounds etc., damage is demonstrable to endothelial cells and the opening of endothelial cellular junctions is seen (28, 29). In the preceding lecture attention was drawn to evidence that stimuli of this kind also evoke a "mesenchymal reaction" in the subendothelial tissues associated with increased production of sulphated glycosaminoglycans and proteoglycans which may serve to entrain TLDL and fibrinogen entering the wall.

Interaction of hyperlipidaemia and altered vascular permeability

Once increased permeability has been established, it can be envisaged that there would be an increased volume of plasma permeating in unit time through a given affected segment of the wall, as well as ingress of the large molecules normally largely excluded. This would permit TLDL and fibrinogen to enter in higher total concentrations at the affected site even in the absence of raised concentrations of these proteins in the circulating plasma, perhaps explaining why atherosclerosis may be initiated and progress in spite of normal or only slightly elevated serum lipid levels in man. However, clearly the process envisaged would be further accelerated with the concomitant occurrence of hyperlipidaemia (hyperlipoproteinaemia).

The interaction between increased vascular permeability and hyperlipidaemia is well illustrated in relation to the pathogenesis of xanthomata (30). These are most commonly seen in man in association with both primary (familial) hyperlipidaemias and the secondary hyperlipidaemias accompanying diabetes mellitus, the nephrotic syndrome, hypothyroidism and obstructive biliary disease. Xanthomata are characteristically distributed at sites subjected to minor trausma (elbows, knees, buttocks); in skin subjected to constant creasing or folding (palmar creases, eyelids); in tendons subjected to friction in the neighbourhood of joints (finger tendons) or in tendons subjected to constant tensional stress (tendo Achillis) If treatment is effective in reducing hyperlipidaemia, xanthomata have been repeatedly noted to decrease in size or to regress, with loss of the lipid contained in them (for references, see Ref.30). On the other hand, in relation to the active progression of lesions in untreated cases of hyperlipidaemia, it was shown by Scott and Winterbourn (31) that when autologous radioactively labelled TLDL was injected into hyperlipidaemic subjects with actively growing xanthomata symmetrically distributed upon both elbows, greater localisation of radioactivity occurred in a lesion subjected to slight but repeated trauma than in a protected lesion on an immobilised limb. These results suggested that xamthomata might arise by leakage of TLDL-rich plasma into the connective tissues, because of localised increase of vascular permeability due to the release (occasioned by repeated slight trauma at the affected sites) of histamine and other vaso-active compounds, with uptake of the lipid (lipoprotein) by tissue histiocytes.

This hypothesis was also supported by experiments in rabbits (30). In this species xanthomata with histological characteristics closely similar to those of human lesions occur spontaneously in animals maintained on lipid-supplemented diets and with hyperlipoproteinaemia. But it was shown that similar lesions could be induced at elective sites by the repeated injection of histamine or bradykinin . No similar lesions could be induced in normolipidaemic animals. Using radioactively labelled rabbit TLDL, or the technique of immunofluorescence, it was shown that the material in the lesions was indeed rabbit TLDL. Moreover it was noted that early induced lesions in animals with moderate hyperlipoproteinaemia regressed when the injections giving rise to altered vascular permeability were interrupted. On the other hand, lesions in animals with gross hyperlipoproteinaemia were more rapidly and easily induced, persisted and became indurated.

Hyperlipidaemia and other risk factors

Epidemiological studies have documented the association of hyperlipidaemia with the various forms of cardiovascular, cerebrovascular and peripheral vascular morbidity and mortality arising from atherosclerosis in the arteries supplying these various territories. Since HDL show relatively little variation in health or disease (9) hyperlipidaemia, in practice, is given expression in elevation of TLDL manifested as increases in one or more of its subclasses.

Moderate hyperlipidaemia. The effects of age, sex, geographic location, cultural conventions etc upon serum lipid levels and the incidence of atherosclerosis on a population basis have been discussed previously (9, 32). Most attention in studies of this kind has been concerned with the quality and proportion of fat in the diet in influencing serum levels. More recently, attention has also turned to the quality of dietary protein (whether from vegetable or animal sources) and the effect of this upon serum lipids and lipoproteins (33, 34).

There is evidence that the apolipoproteins of HDL and of TLDL are synthesized, and the intact lipoproteins assembled, in both the liver and in the epithelial cells of intestinal mucosa (35, 36). Nothing is known of the relative contributions made by each site of synthesis to the total body pools of these lipoproteins, nor is it known how control of the rate of synthesis at these sites is exerted, though clearly one would expect dietary variation to exert effects upon both sites.

Gross hyperlipidaemia. The accelerated rate of development of atherosclerosis, the precocious occurrence of a corneal arcus, and the frequent presence of xanthomata in association with both severe secondary and with primary (familial) hyperlipidaemias is well documented (for references see Ref.37).

Table I

Risk of cardiovascular disease (CVD) and patterns of alterations of lipoproteins in familial hyperlipoproteinaemias. *

Type ≠		Risk of CVD ƚ
I	Gross chylomicronaemia; VLDL, IDL and LDL normal or low; deficient lipoprotein lipases	Normal
IIA	Marked increase of LDL	Very high
IIB	Increased LDL with moderate increase of VLDL	High
III	Increase of LDL and IDL ("floating betalipoprotein")	High
IV	Marked increase of VLDL ± increase of LDL	Increased
V	As for IV, with chylomicronaemia	? Increased

* Slightly modified from Walton (9) with permission of Editor and publishers of Amer. J. Cardiol.
≠ Based on classification of Beaumont et al (38)
ƚ From data of Fredrickson et al (39) and of Slack (40)

In both primary and secondary hyperlipidaemias the high serum TLDL levels reflect a greatly increased total body pool of these lipoproteins with an increase in both intravascular and extra-

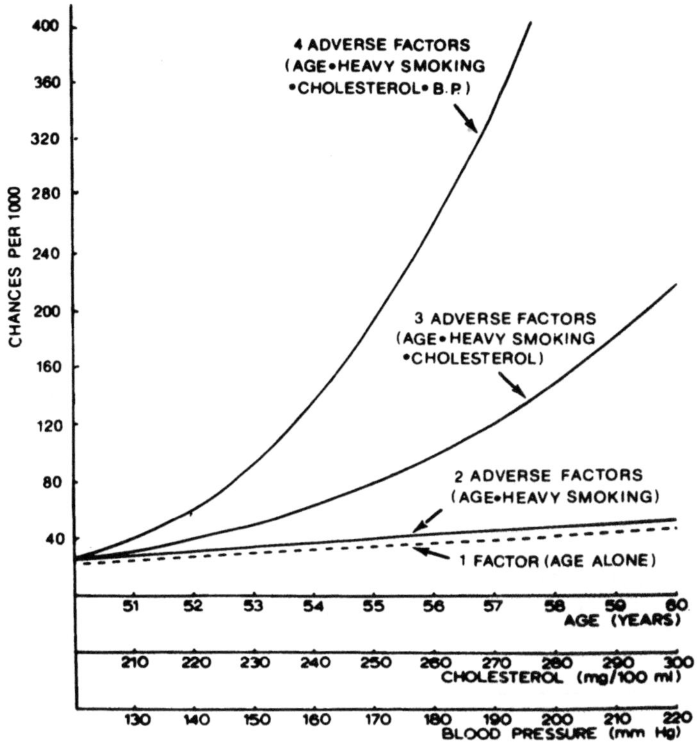

FIGURE. Interaction of risk factors in increasing probability of development of coronary heart disease (C.H.D.) during next 5 years in men aged 50-60 as calculated using multiple logistic equation of Keys et al. (42)

vascular distribution (9). It seems reasonable to infer that the increased extravascular distribution is given expression as TLDL deposits both in the walls of arteries and in the tissues.

Primary hyperlipidaemias classified according to the system proposed by the World Health Organisation (38) show varying risks of development of cardiovascular disease (see Table I).

Consideration of this data allows inferences to be drawn about the relative " atherogenic potential" of lipoproteins of differing molecular size. For example, the rare Type I abnormality due to a gross chylomicronaemia associated with deficiency of adipose tissue or hepatic lipoprotein lipase is not characterised by an accelerated development of atherosclerosis. The

Table II

Possible Mechanisms Whereby "Risk Factors" Exert Their Effect at the Level of the Arterial Wall

Risk Factors	Possible Mechanism
1. Age, smoking (hypoxia), extrinsic or intrinsic vasoactive agents	Increased vascular permeability (increased inflow)+"non-specific mesenchymal response"
2. Age, sex, high fat intake as proportion of total calories	Moderate hyperlipoproteinemia (increased input of total low density lipoprotein even with unchanged volume of inflow.
3. Primary or secondary hyperlipoproteinemia	Gross hyperlipoproteinemia (effect as in 2 but more prounced)
4. Hypertension	Further increase of inflow (effect especially marked in hyperlipidemia)
5. External injury to vessel or coexistent arteritis	Inflammation of arterial wall increasing vascular permeability; obliteration of lymphatic vessels reducing outflow

functional defect in such individuals is an inability to degrade chylomicra into smaller lipoproteins. One might therefore infer that intact chylomicra are too large to enter the arterial wall. At the other end of the scale, it might be expected that the smallest lipoprotein molecules of the TLDL family (the LDL class) might penetrate most easily through the endothelial barrier. This accords with the data of Slack (40) who found that men with the Type IIA or IIB abnormality (preponderant increase of LDL) had an earlier onset of ischaemic heart disease than did men with Types III, IV or V abnormalities. Nevertheless, the incidence rate was high even in persons with a Type IV or V pattern (predominant increase of VLDL). Visual evidence that molecules larger than LDL do, in fact, occur in atherosclerotic lesions was discussed in Lecture I.

Hypertension

Increased blood pressure is generally acknowledged to accelerate arterial disease in man (27) and in experimental animals. The manner in which pressure effects are especially evident at certain sites (for example in the pulmonary artery in pulmonary hypertension, in the proximal segment in coarctation of the aorta, in the mitral and aortic valves and in autologous vein grafts) has been previously discussed (9) and will be dealt with in more detail in Lecture III.

Interaction of hyperlipidaemia and other factors

The manner in which risk factors reinforce one another in a geometric rather than an arithmetic summation of effect has been given mathematical expression. Calculations based on one of the formulae (42) put forward are expressed graphically in the accompanying figure. The manner in which some of the factors probably exert their effect on the arterial wall (summarised in Table II) and summate in so doing, has already been alluded to in this Lecture. Further human and experimental models illustrating the interaction will be discussed in Lecture III.

REFERENCES

1. Pickering, G.W. (1969) Brit.Med.J., 1, 517.
2. Cohen,S.,Freeman,T.and MacFarlane,A.S.(1961) Clin.Sci.,20, 161.
3. Takeda,Y. and Reeve,E.B.(1963) J.Lab.Clin.Med.,61,183.
4. Scott,P.J.and Hurley,P.J.(1969)Israel J.Med.Sci,5,631.
5. Scott,P.J.and Hurley,P.J.(1970) Atherosclerosis,11,77.
6. Walton,K.W.,Scott,P.J.,Dykes,P.W.and Davies,J.W.L.(1965) Clin.Sci.,29,217.
7. Scott,P.J.,Dykes,P.W.,Davis,J.and Walton,K.W.(1963) In Biochemical Problems of Lipids (ed.,Frazer,A.C.) pp.318-324, Amsterdam,Elsevier.
8. Hurley,P.J. and Scott, P.J. (1970) Atherosclerosis, 11, 51.
9. Walton,K.W.(1975) Amer.J.Cardiol., 35, 542.
10. Nagy,Z.(1974) In Arterial Lesions and Arteriosclerosis (ed.Jellinek,H.) pp.110-137, London, Plenum Press.
11. Adams, C.W.M.(1967) Vascular Histochemistry in Relation to the Chemical and Structural Pathology of Cardiovascular Disease, p.74,London, Lloyd-Luke .
12. Constantinides,P.(1965) Experimental Atherosclerosis, pp.27-51, Amsterdam, Elsevier.
13. Björkerud,S. and Bondjers,G.(1973) Nutr.Metab. 15, 27.
14. Minick, C.R., Murphy, G.E. and Campbell,W.G.(1966) J.Exp.Med., 124, 635.
15. Minick, C.R.,(1976) Ann.N.Y.Acad.Sci., 275, 210.
16. Aok-Upmark,E. and Fajers,C-M.,(1956) Acta Med.Scand. 155, 275.
17. Shrire,V. and Anderson,R.A.(1964) Quart.J.Med.,33,439.
18. Turnbull, H.M.(1915) Quart.J.Med., 8, 201.
19. Nichols, E.F. (1940) Am.Intern.Med., 14, 960.
20. Walton,K.W. (1974) Adv.Drug Research, 9, 55.
21. Shimamoto, T. (1974) In Atherosclerosis III. (eds.Schetter,G. and Weizel,A.) pp.68-82, Berlin, Springer Verlag.
22. Hueper,W.C.(1944) Arch.Pathol.38, 162,245,350; (1945) Arch. Pathol., 39, 57, 117, 187.

23. Hueper W.C. (1956) Amer.J.Clin.Pathol., 26, 559
24. Myaskinov, A.L. (1958) Circulation, 17, 99.
25. Astrup, P., Kjeldsen,K and Wastrup,J. (1967) J.Atheroscler. Res., 7, 343.
26. Kjeldsen,K. In Atherosclerosis:Proceedings of the Second International Symposium (ed.Jones R.J.) pp.378-381, Berlin, Springer Verlag.
27. Dawber,T.R.,Kannel,W.B.,Revotskie, N. and Kagan, A. (1962) Proc.R.Soc.Med.,55, 265.
28. Constantinides, P. and Robinson,M. (1969), 88, 99, 106, 113.
29. Kerenyi, T. and Hüttner, I. (1974) In Arterial Lesions and Arteriosclerosis (ed.Jellinek,H.) pp.91-102, London, Plenum Press.
30. Walton,K.W., Thomas C. and Dunkerley,D.J. (1973) J.Pathol.,109, 271.
31. Scott, P.J. and Winterbourn, C.C. (1967) J. Atheroscler.Res. 7, 207.
32, Walton, K.W. (1969) In The Biological Basis of Medicine, (eds.Bittar,E. and Bittar,N.) Vol.6, pp.193-223, New York, Academic Press.
33. Hamilton,R.M.G. and Carroll,K.K. (1972) Atherosclerosis, 24,47.
34. Sirtori, C.R., Agradi, E., Conti,F., Mantero, O. and Gatti, E. (1977). Lancet, 1, 275.
35. Hamilton,R.H. (1972) Adv.Exper.Med.Biol., 26, 7.
36. Walton, K.W., Morris, C.J. and Bradby, G.V.H. (1977) Nature (Lond.) in press.
37. Walton, K.W. In Textbook of Geriatric Medicine and Gerontology, (ed., Brocklehurst,J.C.) pp.77-112, Edinburgh, Churchill Livingstone.
38. Beaumont, J.L., Carlson, L.A., Cooper, G.R., Fejfar, Z., Fredrickson, D.S. and Strasser, T. (1970) Bull.Wld.Hlth.Org., 43, 891.
39. Fredrickson, D.S., Levy, R.I. and Lees, R.S. (1967) New Eng. J.Med., 276.
40. Slack, J. (1969) Lancet, 2, 138¼.
41. Hüttner, I. (1974) In Arterial Lesions and Arteriosclerosis (ed.Jellinek, H.) pp.128-134, London, Plenum Press.
42. Keys,A., Aravanis, C., Blackburn, H., Buchem, F.S.P.van, Buzina,R., Djordjevic, B.S., Fidanza, F., Karvonen, M.J., Menotti, A., Puddu, V. and Taylor, H.L. Circulation, 1972, 45, 815

MODELS FOR THE STUDY OF TISSUE DEPOSITION AND SYNTHESIS OF LIPOPROTEINS (LECTURE III)

K.W. Walton

Department of Experimental Pathology

University of Birmingham, England

Experimental animal models

Efforts have been made, using a wide variety of animal species, to produce an acceptable animal model of human atherosclerosis. No single model is accepted universally as reproducing all aspects of the condition in humans (for review, see Ref.1) although primate models, and especially those involving the apes, such as the chimpanzees, probably approach this ideal most closely because both the lipid composition (2) and the immunological characteristics (3) of their lipoproteins are very similar to those of man.

However, because of the expense involved in obtaining and maintaining primates and also because of restrictions imposed by the Governments of some countries on the use of certain genera of primates, many laboratories have perforce to work with cheaper and more convenient laboratory animals. It is now becoming recognized that certain species in which the principal lipid-carrying protein is a high-density lipoprotein (dogs, rats, hamsters) are resistant to the induction of experimental atherosclerosis. On the other hand, in species in which the principal lipid-carrying proteins are low-density lipoproteins (as in man) dietary manipulation can be shown to increase the level of these lipoproteins in the blood and to be attended by the development of arterial lesions. Among this latter group is the rabbit. From the limited view-point of examining, under experimental conditions, whether an insudative mechanism operates in atherogenesis in species other than man, it is worthwhile to re-evaluate the validity of use of the lipid-fed rabbit as an experimental model.

Experimental atherosclerosis in the rabbit

It has been shown that whether rabbits are maintained on cholesterol-supplemented diets, or on semi-synthetic diets containing fat (e.g. beef lard) without added cholesterol, the animals develop not only a hypercholesterolaemia, but also a hyperlipoproteinaemia characterised by an increase mainly of TLDL. The distribution of density sub-classes within the TLDL varies with the nature of the diet. In the cholesterol-fed animal there is a marked increase of a lipoprotein with VLDL-like density-characteristics but which differs from normal rabbit VLDL in containing a much higher proportion of cholesterol than normal (4). The protein moiety of this 'abnormal' lipoprotein also differs from normal rabbit VLDL in containing an unusually high proportion of a peptide probably analgous with human arginine-rich peptide (apolipoprotein E) according to some authors (5 - 7) and in its large and relatively uniform size as visualized by electron microscopy (7). Cholesterol-fed animals often show a reciprocal reduction in high-density lipoproteins (4). On the other hand, beef fat fed rabbits show a more moderate hyperlipidaemia, attended by an increase of both HDL and TLDL but the latter show a distribution of density-classes within the TLDL which more closely approximates to that in normal rabbits (4).

Despite these differences, it was established in animals on both the above dietary regimes that the arterial lesions, when examined by the technique of immunofluorescence, showed the presence of rabbit TLDL (8) distributed in a manner closely similar to that previously found in human lesions using a similar technique (9). Moreover, when isotopically-labelled rabbit TLDL was injected intravenously, the labelled molecule was found to localise in the aorta and other lesions in these animals (4). This model was found also to be suitable for study of the mechanism involved in the deposition of lipid (lipoprotein) into certain intravascular sites to give rise to lesions which have long been known to be broadly correlated with the occurrence of human atherosclerosis, namely xanthomatosis and corneal arcus formation (4.10,11).

The Dietary factor in atherosclerosis

In the rabbit model, hyperlipidaemia is induced solely by dietary manipulation. At the height of the hyperlipidaemia so induced there is evidence that the synthesis of cholesterol and phospholipids by the liver is greatly reduced (for references see Ref.12) and that synthesis of lipoproteins by parenchymal liver cells is similarly reduced (8). It may be inferred therefore that the great bulk of the circulating lipoproteins is of intestinal origin.

In man, epidemiological observations suggest that the mean

level of circulating lipids and lipoproteins is influenced by the proportion of the dietary intake which is in the form of fat (13) and, in particular, of saturated fat. But in normal man, unlike the rabbit, there appears to be a control mechanism, safeguarding against grossly excessive fat absorption from the gut since, following a very large fat-meal, only a proportion of the fat is absorbed, the remainder being excreted as faecal fat.

While there is good evidence for the transformation of absorbed fat into lipoprotein by intestinal epithelial cells in the rat (14) there has hitherto been little direct evidence that intestinal lipoprotein synthesis also occurs in man. Recent work by the author and his collaborators (15) using immunofluorescence established that apolipoproteins A, B, C and E were all demonstrable both in liver cells and in intestinal epithelial cells from normal subjects. Ultrastructrual studies, using the immunoperoxidase technique and a labelled antiserum to apolipoprotein B suggested that in intestinal epithelial cells the smooth endoplasmic reticulum, the Golgi apparatus and secretory vesicles derived from the latter, served as a pathway for the assembly, transport and secretion of the serum lipoproteins into intestinal lymph. An essentially similar intracellular pathway has been described for liver cells (16).

Currently virtually nothing is known of the relative contributions made by liver and intestinal synthesis to the total body pool and the circulating level of lipoproteins in man. But it is possible that in certain human pathological conditions (perhaps particularly primary hypertriglyceridaemias characterised by an increase of endogenous VLDL) the contribution by the intestinal site may be excessive. Hyperlipidaemias of this kind are relatively common and associated with an increased incidence of atherosclerosis (17).

Contributory factors accelerating insudation of lipoproteins

Increased blood pressure has been clearly established as a factor accelerating arterial disease in man (18) and in experimental animals (19). It can be envisaged that, once selective alteration of permeability (a localised "leak") has been established in an artery, the volume of plasma transuding through the affected segment would be increased in unit time. In this way the total concentration of potentially reactive molecules (TLDL and fibrinogen) entering the segment in unit time would be increased even if the plasma levels of these components were normal. The opportunity for their entrapment in the gel would be thus increased. But clearly the effect would be greatly magnified by concomitant hyperlipidaemia. This mechanism can be evoked to explain the clinical observation that hypertension increases the risk of development of cardiovascular disease but that the risk is magnified in subjects with elevated levels of blood lipids.

The effect is particularly marked at certain sites. For example, atherosclerosis of the pulmonary artery in man is only seen in association with pulmonary hypertension (20) and in co-arctation of the aorta is sharply delimited to the proximal (high-pressure) segment. Attention has previously been drawn (21) to the significance of atherosclerosis of the heart valves. This is seen only in the valves in the left side of the heart (mitral and aortic valves) which are at arterial pressure and not in the valves of the right side of the heart which are at venous pressure. Moreover the distribution of lipid (lipoprotein) infiltration is of significance. In the mitral valve, lipoprotein infiltration is confined to the ventricular half of the valve (i.e., to the surface exposed to high pressure during closure of the valve with systole). In the aortic valve, lipoprotein infiltration is maximal at the base of the pocket formed by each cusp (i.e. to the point of maximal pressure from backflow during diastole).

While atherosclerosis is normally found only in arteries (high pressure vessels) and not in veins (low pressure vessels), when autologous vein grafts are used to bypass occluded vessels and become exposed to arterial pressure, lipoprotein infiltration, closely resembling that seen in atherosclerosis of arteries, can be shown to occur (22). Contrariwise, arteries normally exposed to systemic pressure do not develop atherosclerosis if, owing to developmental abnormalities, they are exposed only to lower pressures. This is well seen in the case of the rare developmental abnormality affecting the left descending coronary artery in which this vessel arises from the pulmonary trunk instead of the aorta (23).

Summary of current views of atherogenesis

It is submitted that the evidence discussed in Lecture I is overwhelmingly in favour of the lipid present in atherosclerotic lesions arising from 'insudation' of TLDL into the arterial wall rather than from 'fatty degeneration' of intrinsic components of the wall, or from breakdown of the components of thrombi deposited on the surface and incorporated into the wall.

In Lecture II, it was proposed that increase of the relative permeability of the arterial wall by TLDL is probably the source of origin of atherosclerotic lesions and probably also of certain extravascular lesions broadly associated with atherosclerosis. Localised alteration of permeability is probably the determinant of the topographic distribution of lesions but diffuse increase in vascular permeability increases the severity of lesions. Since the arterial wall is living connective tissue, it reacts at sites of altered permeability by increased mesenchymal activity. This gives rise to increased synthesis of components of the ground substance which entrap and give rise to firm binding of TLDL and fibrinogen. Hyperlipidaemia increases the rate and severity of this process but is by no means the only determinant since hypertension, changes in the arterial wall caused by trauma, arterial disease etc., also contribute. The available evidence suggests that where several factors operate, they act synergistically to produce a summation which is geometric rather than simply arithmetic.

In Lecture III it was shown that animal models, including the classic but disputed model of the lipid-fed rabbit, can be used to validate some of the views expressed above. Evidence was also presented for lipoprotein synthesis in man both by the liver and by the intestine and attention was drawn to the need to define the contribution made by each site of synthesis to the total body pool of TLDL and thus to hyperlipidaemia and the increased risk of atherosclerosis. It was also shown that certain aspects of atherogenesis can be illuminated by examining sites other than the arterial plaque itself, in man,and this was illustrated by a number of examples relating to the effect of increased blood pressure.

REFERENCES

1. Wissler, R.W. and Vesselinovitch, D. (1974) In Atherosclerosis III (eds. Schettler, G. and Weizel, A.) pp. 319-325, Springer Verlag, Berlin.

2. Blaton, V. and Peeters, H. (1976) Adv. Exper. Med. Biol., 67, 33.

3. Walton, K.W., Hitchens, J., Valente, A.J. and Blaton, V. (1977) Nature (Lond.) - in press.

4. Walton K.W., Thomas C. and Dunkerley, D.J. (1973) J. Pathol., 109, 271.

5. Shore, B. and Shore V. (1960) J. Lipid Res., 1, 321.

6. Camejo, G., Bosch, V., Arreaza, C. and Mendez, H.C., (1973) J. Lipid Res., 14, 61.

7. Camejo, G., Bosch, V. and Lopez, A. (1974) Atherosclerosis, 19, 139.

8. Walton, K.W., Dunkerley, D.J., Johnson, A.G., Khan, M.K., Morris, C.J. and Watts, R.B. (1976) Atherosclerosis, 23. 117.

9. Walton, K.W. and Williamson, N. (1968)J. Atheroscler.Res.8,599.

10. Walton, K.W. (1973) Nutr. Metab., 15, 37.

11. Walton, K.W. and Dunkerley, D.J. (1974) J. Pathol., 114, 217.

12. Shore B. and Shore V. (1976) Adv. Exper.Med. Biol., 67, 123.

13. Keys, A. (1963) In Atherosclerosis and its Origin (Ed. Sandler, M. and Bourne, G.H.) pp.263-299. Academic Press, New York.

14. Windmueller, H.A. and Spaeth, A.E. (1972) J. Lipid Res.,13, 92.

15. Walton, K.W., Morris, C.J., Bradby, G.V.H. and Pagnan, A. (1977) Nature (Lond) - in press

16. Hamilton, R.H. (1972) Adv. Exper. Med. Biol., 26, 7.

17. Carlson, L.A. and Böttiger, L.E. (1972) Lancet, 1, 865

18. Dawber, T.R., Kannel, W.B., Revotskie, N. and Kagan, A. (1962) Proc. R. Soc. Med., 55, 265.

19. Heptinstall, R.H., Barkley, H. and Porter, K.A. (1958) Angiology, 9, 84.

20. Heath, D., Wood, E.H., Dushane, J.W. and Edwards, J.E. (1960) Lab. Invest., 9, 259.

21. Walton, K.W., Williamson, N. and Johnson, A.G. (1970) J. Pathol., 101, 205.

22. Walton, K.W., Slaney, O.F. and Ashton, F. - in preparation.

23. Burch, G.E. and De Pasquale, N.P. (1964) Amer. J. Med., 37, 159.

DIETARY EFFECTS ON LIPOPROTEIN COMPOSITION

Cesare R. Sirtori, Gian Carlo Ghiselli and
M. Rosa Lovati

Center E. Grossi Paoletti for the Study of
Metabolic Disease and Hyperlipidemias, University
of Milano - 20129 Milano, Italy

INTRODUCTION

Qualitative and quantitative dietary changes can markedly affect plasma lipid levels. Variations of lipidemia have an obvious counterpart in changes of lipoprotein levels and composition. In spite of the large number of studies evaluating dietary effects on plasma lipid levels, clinical studies on the relative composition and structure of lipoproteins following dietary changes are relatively scarce. Numerous studies report, on the other hand, changes of lipoprotein composition and structure in animals fed different diets; this is particularly true for diets used with the purpose of inducing experimental atherosclerosis.

This report will analyze available information on lipoprotein composition following dietary changes, expecially aimed to modify plasma triglyceride and cholesterol levels. In both cases, a separate consideration is given to diets which induce increases of plasma triglyceride and/or cholesterol, and to diets which reverse these situations in clinical syndromes.

LIPOPROTEIN CHANGES IN DIET INDUCED HYPERTRIGLYCERIDEMIAS

Increases of triglycerides (TG) and particularly of TG rich lipoprotein fractions, i.e. chylomicrons and very low density lipoproteins (VLDL) are induced by diets exceedingly rich in calories, carbohydrates (1), ethanol (2) and by combinations of these. Addition of fat to ethanol or to carbohydrates worsens diet induced hypertriglyceridemias.

Carbohydrates

Hypertriglyceridemias, particularly type IV hyperlipoproteinemia, may be secondary to diets containing an excessive amount of carbohydrates (CHO)(1). This situation may be reproduced in animals and in man. The resulting clinical syndrome is similar to that found in patients with hyperlipoproteinemias.

After induction with CHO rich diets, the resulting excess of plasma TG is circulated as part of the VLDL molecules,assembled and secreted primarily by the liver. Studies on the mechanism of the changes in plasma TG transport following CHO induction, have demonstrated a significantly altered VLDL ($d<1.006$) composition in normal volunteers and patients with hyperlipidemia (3). Switching from a diet containing 45% of CHO with 15% of protein, to a diet containing 80% CHO and 20% protein, results in all cases in increased plasma TG, on the average about doubled, without significant changes of plasma cholesterol levels. TG content of VLDL (Table 1) is about triplicated and the ratio of TG to protein more than doubled. These results are consistent with the hypothesis (4) that the liver may change the pattern of secretion of TG rich particles following CHO induction. They also indicate that VLDL composition may significantly differ relative to the dietary habits and particularly to the CHO intake.

The apoprotein composition of VLDL, according to Carlson and Ballantyne (5), may be altered in the presence of an increased TG/protein ratio in VLDL particles. In particular, significant changes may be observed in the relative proportion of apoproteins CII and CIII. In case of markedly hypertriglyceridemic VLDL (i.e.,in type V hyperlipoproteinemia) an almost total absence of CII is found, whereas CII normally prevails over CIII. These observations confirm the hypothetical role of CII in activating lipoprotein lipase (6).

After CHO induction in normal man, Schonfeld et al. (7) recently described, however, increased CII and decreased CIII 2 in both chylomicrons and VLDL. Apo Al was markedly decreased in HDL, which also become en-

Table I

CHANGES IN VLDL COMPOSITION FOLLOWING CARBOHYDRATE INDUCTION IN MAN (3).

Diet:	*CHO 45%* *Prot. 15%*	*CHO 80%* *Prot. 20%*
	PLASMA	
$\overline{X}$ TG	214.7 mg/dl	534.2
$\overline{X}$ Chol.	235.2	235.3
preβ	34.6%	61.7
	VLDL (d < 1.006)	
$\overline{X}$ TG	137.4 mg/dl	438.0
$\overline{X}$ Chol.	31.1	74.2
Prot.	44.4	62.1
TG/P	3.09	7.05
C/P	0.70	1.19
TG/C	4.40 (3-6.7)	5.90 (4.4-7.5)

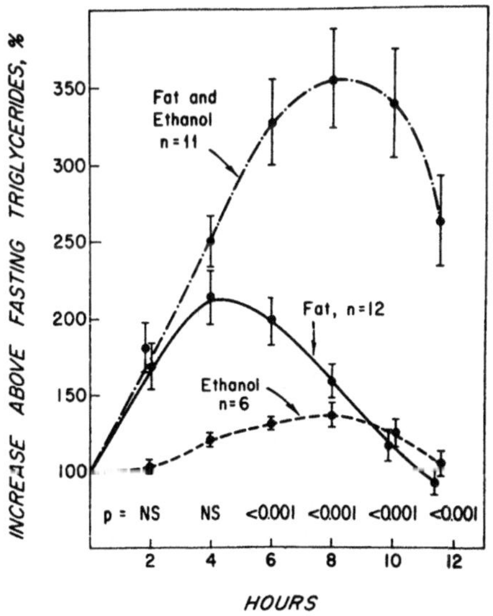

Figure 1: Plasma triglyceride changes following ethanol, fat (100 g of corn oil) and fat + ethanol in healthy volunteers (2).

riched with triglycerides. In normals, therefore, apoprotein changes are indicative of an improved metabolic clearance; it is possible, on the other hand, that in patients with hypertriglyceridemia, CHO induction may cause secretion of VLDL particles with reduced activator apoproteins.

Lipoprotein changes following CHO induction, in conclusion, are characterized by TG enrichment of VLDL and possibly other lipoprotein fractions, with apoprotein changes normally consistent with an improved metabolic clearance of the newly secreted lipoproteins. Data on apoprotein composition in patients with hypertriglyceridemia, suggests that the situation may be different in these. Thus, the test of CHO induction may be a useful tool for the diagnosis of the dietary origin of hypertriglyceridemias.

Ethanol

Ethanol represents a very potent inducer of hypertriglyceridemia. This effect of ethanol may be particularly striking in selected patients (8), predisposed to this type of hypertriglyceridemia, but may be found in practically all individuals following a load of ethanol (9).

Several studies have dealt with lipoprotein composition in ethanol-induced hypertriglyceridemia. Wilson et al. (2) showed that ethanol administration to volunteers induces a 50% rise of plasma TG, and that additional fat (100 g of corn oil) causes a dramatic increase of triglyceridemia (Fig. 1). No significant changes of postheparin lipolytic activity (PHLA) and of plasma glucose and insulin are observed, whereas both dietary treatments, alone and in combination, significantly increase plasma FFA.

Analysis of plasma lipoproteins in patients following fat and ethanol, demonstrates significant increases of both chylomicrons (Sf > 400) and VLDL (Sf 20-400). Subfractionation of lipoproteins by analytical ultracentrifuge shows that (Tab. 2) lipoproteins of Sf 100-400 and 20-100 are significantly increased by ethanol and fat, whereas lipoproteins Sf 12-20 (corresponding to IDL) and Sf 0-12 (corresponding to LDL) are not modified. An interesting observation is the significant increase of HDL following this dietary regimen.

The tentative conclusive of this study was that lipoproteins increased following fat and ethanol are

Table II

LIPOPROTEIN CONCENTRATIONS AFTER INGESTION OF FAT AND ETHANOL (2)[+]

Time after ingestion	Lipoprotein concentration (mg/100 ml)				
	$S_{f\,100-400}$	$S_{f\,20-100}$	$S_{f\,12-20}$	$S_{f\,0-12}$	HDL
CONTROL	21.2	46.6	18.6	418	111
4 HOURS	61.5	63.9	15.0	450	149
10 HOURS	58.1	87.2	17.8	372	167

+ Analytical ultracentrifugation of plasma at density 1.20 g/ml. Lipoprotein classes are expressed in S_f 1.063 units.

likely to be all of liver origin, independent of changes in PHLA (10). This conclusion was, however, not supported by the fatty acid composition of VLDL after fat and ethanol, similar to that occurring after fat alone, and by the ineffectiveness of medium chain triglycerides in modifying ethanol induced hypertriglyceridemia, also suggesting a possible impairment of clearing mechanisms.

Somewhat different lipoprotein changes were described by Avogaro and Cazzolato (11) in alcoholic subjects given a large load of ethanol (180 g in 6 hrs.) and studied for the following 24 hrs. The acute hypertriglyceridemia (from 26% to 367% above basal values) was secondary to TG increases in all lipoprotein fractions, maximal in VLDL, followed by HDL and LDL. Analyses of lipoprotein distribution by polyacrylamide gel electrophoresis and analytical ultracentrifuge were suggestive of the formation of an extra-lipoprotein fraction, presumably of d 1.006-1.019 (IDL). Detection of this lipoprotein fraction was variable, although rather constant, among subjects. It was often found to be associated with the appearance of a "pre-β" band in the electrophoresis of LDL-HDL ("sinking pre-β") separated by preparative ultracentrifuge. This last data is somewhat in contrast to that of Wilson et al. (2) in patients given ethanol and fat. It is possible that when ethanol is given alone in toxic doses it may increase lipoproteins of intermediate density, whereas when given together with fat, only physiological lipoproteins of very low density are increased. Further apoprotein studies are needed to detect specific apoprotein components following ethanol.

In conclusion, ethanol administration increases VLDL and chylomicrons and determines a general TG enrichment in all lipoproteins. There is disagreement between data reporting no specific changes of lipoprotein distribution, and other describing increases of IDL. An increase of HDL following ethanol, has been reported by both Wilson et al. (2) and by Avogaro (12). This finding may cast some doubts on the role of ethanol in atherogenesis. Whereas, in fact, ethanol is generally considered an atherogenic agent (13), increases of HDL may provide a protective effect (14).

HYPOTRIGLYCERIDEMIC DIETS AND LIPOPROTEIN COMPOSITION AND STRUCTURE

The most common dietary treatment for hypertriglyceridemias is CHO reduction. CHO restricted diets, similar to those commonly used in therapy of diabetes mellitus (15), are generally effective on plasma TG levels, but have been shown to elicit some yet unexplained lipoprotein modifications. In particular, as also observed after treatments with hypotriglyceridemic drugs (e.g., clofibrate or nicotinic acid), an inverse change between VLDL and LDL may be detected (16). In some subjects significant increases of LDL cholesterol may occur following CHO restriction. Carlson et al. (17) proposed mathematical formulas, to predict changes of LDL cholesterol following diets, or drugs of the clofibrate type.

A definitive explanation for these findings, rather disturbing, in view of the higher atherogenicity of LDL as compared to VLDL, is yet to be found. According to Eisenberg (18), three possibilities should be considered: a) increased synthesis of apo B and secretion of a larger number of VLDL particles; b) decreased rate of removal of intermediate lipoprotein particles from the circulation; and c) decreased catabolism of LDL.

Although none of these three hypothetical mechanisms has been analysed in detail, in a recent study by Vessby and Lithell (19), patients with hypertriglyceridemia (particularly of type IV), treated by CHO restriction, could be subdivided into two groups: a) those showing an increase of LDL cholesterol and b) those showing a decrease, the cut-point between the two being at an LDL-cholesterol level of approximately 150 mg/dl (Fig. 2).

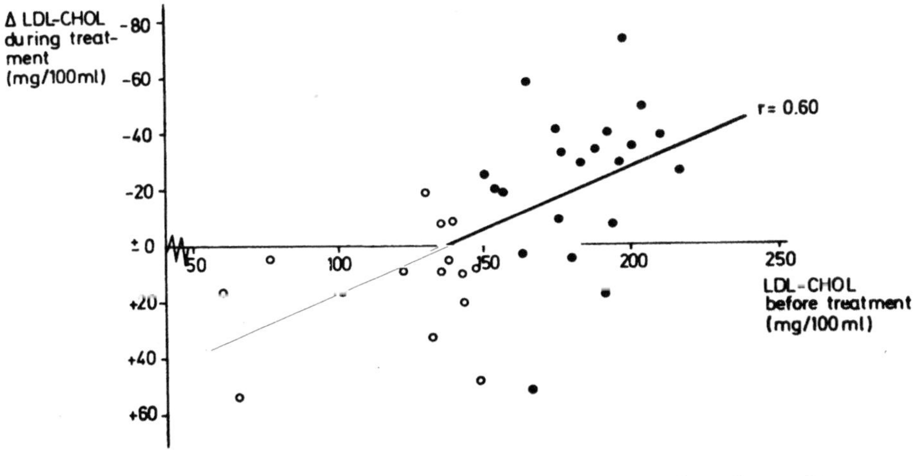

Figure 2: Variations of LDL cholesterolemia after carbohydrate restriction in type IV patients (19).

Subjects with LDL cholesterol above 150 mg/dl generally had a decrease after CHO restriction, whereas those starting from a lower level mostly had an increase. Analysis of the lipoprotein composition of the two groups of patients showed a different cholesterol/triglyceride (C/TG) ratio in LDL and HDL. In particular, the C/TG ratios were lower in patients responding to therapy with an elevation of LDL cholesterol. These patients also had significantly lower HDL cholesterol levels. After CHO restriction,patients responding with increased LDL cholesterol had increased C/TG ratio in LDL, whereas no significant changes were noted in HDL composition.

According to Vessby and Lithell this data confirms the hypothesis that there may be different subclasses of type IV patients. Some patients have an increased number of VLDL particles of relatively normal composition and

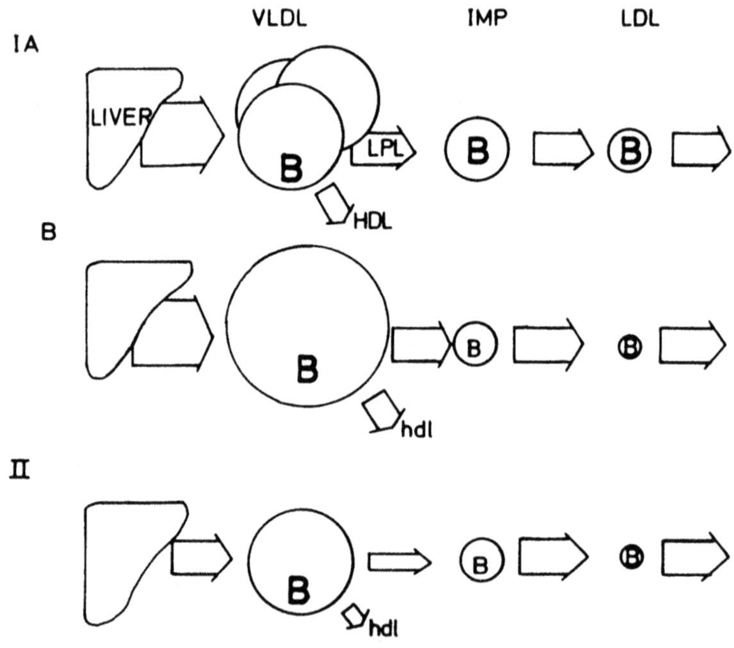

Figure 3. Possible mechanisms of VLDL catabolism in type IV patients after carbohydrate restriction.
IA - Normal VLDL composition and clearance (in some cases leading to increased LDL cholesterol).
IB - VLDL of large size and decreased clearing capacity as witnessed by low HDL.
II - Impaired removal capacity for VLDL (19).

size: these patients have an adequate clearing capacity, sufficient to convert a normal number of VLDL particles to LDL, every particle containing a normal amount of apoprotein C in relation to apo B. It is possible that some of these patients may have elevation of LDL cholesterol following the diet, if LDL clearance is impaired. In contrast, in other patients, an accelerated VLDL synthesis, generally with increased TG content per particle may be present; CHO restriction in these may induce decreased TG synthesis, thus improving clearance and leading, eventually, to increased LDL cholesterol levels. This hypothesis is supported by the finding of low HDL levels, consistent with reduced C proteins. In other words, the final catabolic step of TG rich lipoproteins, i.e. LDL removal, may become saturated, both when VLDL catabolism is enhanced and, paradoxically, even when VLDL triglyceride synthesis is reduced (Fig. 3).

EFFECT OF DIETARY FATTY ACIDS ON LIPOPROTEIN COMPOSITION

Changes of dietary fatty acids, particularly increases of the polyunsaturated/saturated (P/S) ratio, are widely used treatments for hyperlipidemias and expecially for hypercholesterolemias (15). The mechanism of the hypocholesterolemic effect of polyunsaturated (PU) fatty acids is still the object of dispute. Studies on the composition and structure of lipoproteins following diets with different fatty acid compositions, both in man and in experimental animals, have however provided findings which may help explain the cholesterol lowering and possibility anti-atherosclerotic effects of PU fatty acids.

Administration of diet rich in PU fatty acids (i.e., daily consumption of a large dose of a vegetable oil) may lead to rapid changes in the composition of plasma fatty acids. A doubling of the percentage of linoleate may be observed after addition of 100 ml of sunflower oil to the diet for just one day (20). Long-term administration of diets rich in linoleate may lead to marked enrichment of this fatty acid in the plasma cholesterol ester fraction (21), thus, according to Kingsbury (22), exerting a protective effect against atherosclerosis.

Studies on the effect of different fatty acids on lipoprotein secretion have been reported by Ockner et al. (23). These Authors examined the composition and distribution of lymph lipoproteins following duodenal instillation in the rat of various long chain fatty acids in the form of mixed fatty acid-monoolein-taurocholate micelles. All fatty acids (saturated and unsaturated) cause significant increases of chylomicron triglycerides. However, a significant difference may be observed between the effect of palmitate-monoolein micelles and that of oleate-monoolein or linoleate-monoolein micelles. Whereas, in fact, after palmitate lymph cholesterol is equally distributed between chylomicrons and VLDL (Table 3);after oleate and linoleate, the largest portion of lymph cholesterol is found in chylomicrons and an approximately equal percentage in VLDL and d > 1.006 lipoproteins. Administration of progressively larger amounts of linoleate-monoolein micelles, in particular, results in a linear increase of chylomicron lipids and no significant changes of VLDL, whereas after palmitate-monoolein micelles both chylomicron and VLDL fractions are lenearly increased. These observations are of particular interest, since the half--life of chylomicrons in rats is approximately 10 minutes, versus 21 minutes for VLDL, thus suggesting an easier clearing of triglyceride-rich particles after monounsaturated and polyunsaturated fatty acid feeding. These findings suggest changes in the physico-chemical properties of lipoproteins when the diet is enriched in unsaturated fatty acids. According to the model described by Spritz and Mischkel (24), saturated and unsaturated fatty acids differ in molecular configuration and in transition point and are possibly absorbed over a different length of intestine. Oleic and linoleic acid, absorbed over a shorter intestinal segment,would produce a smaller number of large transport particles, whereas palmitate, absorbed over a longer segment of intestine, would develop smaller particles (23).

Studies in rabbits, fed a cholesterol rich diet, and the same diet with the addition of corn oil (a PU rich oil) or coconut oil (rich in saturated fat), were carried out by Stange et al. (25). These authors noted significant reduction of atheromatosis in the corn oil fed grup, in spite of insignificant changes of plasma cholesterol levels. HDL cholesterol was increased in the corn oil treated group, as compared to the others, and interesting differences in the electron-microscopical features were reported. In particular, in the cholesterol and coconut oil-cholesterol fed groups, there was frequent stacking and fusion of the lipoprotein particles, with formation of rosette-like structures. These were not detected in the control and cholesterol-corn oil fed groups, where particles were of regular size and with rare fusions. Similar findings have been reported by Rodriguez et al. (26), who noted, after cholesterol feeding, an enrichment

Table III

CHOLESTEROL DISTRIBUTION IN VLDL AND CHYLOMICRONS AFTER DUODENAL INSTILLATION OF DIFFERENT FATTY ACID EMULSIONS (23)

Duodenal Infusion	CHOLESTEROL DISTRIBUTION		
	Chylos	VLDL	d > 1.006
NaCl	16.8 %	53.9	29.3
16.0	40.3	40.3	19.5
18.1	54.9	25.9	20.0
18.2	53.7	25.5	20.9

Half-life of VLDL : 21.1 ± 7.4 min
" " " Chylos: 10.3 ± 2.6 "

of 18:1 in the VLDL of cholesterol-treated rabbits (VLDL is the major form of cholesterol transport in these animals). The rosette-like formations and stacks were correlated by these Authors to the high content of sphingomyelin in cholesterol-rich lipoproteins (27).

A triglyceride lowering effect has been also described for diets rich in PU fatty acids. Chait et al. (28) noted a triglyceride decrease of 35% when a high P/S diet is substituted to a common saturated fat diet. This effect is found both in normal subjects and in patients with hypertriglyceridemia. Triglyceride reduction occurs mainly in the VLDL fraction; during the saturated fat diet the triglyceride/cholesterol ratio in VLDL is 2.7, whereas after PU fat it decreases to 2.17. During the PU diet there is probably a reduction of VLDL synthesis. This is shown by the lower (about 60%) incorporation of labelled linoleate in VLDL-TG as compared to labelled palmitate. This data, partially confirming the conclusions of the study by Ockner et al. (23) in rats, supports the concept that there is no necessity to use different diets for hypertriglyceridemic and hypercholesterolemic patients. In fact, just one single diet, with a high P/S ratio, may be useful in both syndromes.

Physico-chemical properties of lipoproteins following diets with high a P/S ratio have not been the object of detailed investigations. Some data are reported in the studies by Peeters and Blaton on the effect of polyenyl phospholipids, presented in another paper in this course. Preliminary findings by Taunton et al. (29) have, however, indicated that diets with a very high P/S ratio significantly increase fluidity of lipoproteins (as measured by the motional properties of spin label TEMPO and by EPR of fatty acids). Lipid fluidity is insensitive to temperature and is normally maximal in VLDL as compared to LDL and HDL. After a diet with a high P/S ratio, fluidity of VLDL and LDL is significantly increased, whereas that of HDL is unmodified. The effects of dietary PU fatty acids on lipoprotein metabolism may, therefore, be due in part to increased fluidity of lipoprotein lipids.

DIETS AFFECTING PLASMA CHOLESTEROL LEVELS

Experimental induction of hypercholesterolemia

Dietary induction of hypercholesterolemia has been described very early in the study of atherosclerosis, and has provided some of the best models for research on atherogenesis and for evaluation of pharmacological, surgical and dietary means to reduce atheromatosis. Lipoprotein changes in some of the models which have received the most attention will be described here.

Rabbit - The earliest description of dietary induction of hypercholesterolemia in animals has been that of the cholesterol fed rabbit (30). Cholesterol feeding (1-2 g/day) rapidly induces a dramatic hypercholesterolemia in the animal, with the development of atheromatous lesions within a few weeks of treatment. Some variability of different rabbit strains in the atheromatous response has been reported (31).

Analysis of plasma lipoproteins following cholesterol induction in rabbits has shown an enrichment of cholesterol esters in lipoproteins of density less than 1.019 (VLDL and IDL) (Table 4). Cholesterol esters, normally about 5% of d < 1.019 lipoproteins, become 40-60% of these lipoproteins (26). Change in cholesterol distribution is gradual with time, as described by Brattsand (32). This Author suggests that administration of semisynthetic diets containing coconut oil or butter and a relatively low percentage of cholesterol, may induce (at plasma cholesterol concentrations of 250-400 mg/dl) a hyper-LDL cholesterolemia, similar to type II disease in man. HDL lipoproteins, markedly decreased by diets of this type, become

Table IV

LIPOPROTEIN DISTRIBUTION IN CONTROL AND HYPERCHOLESTEREMIC RABBITS

(mg/dl; $\bar{X} \pm$ SD of six rabbits)

	CONTROLS	HYPER CHOL.
VLDL	0.4 ± 0.1 (15.3%)	19.7 ± 1.5 (46.3)
IDL	0.4 ± 0.1 (15.6)	13.2 ± 1.0 (30.9)
LDL	0.7 ± 0.1 (26.8)	6.5 ± 0.1 (15.2)
HDL	1.1 ± 0.2 (42.3)	3.2 ± 0.1 (7.5)

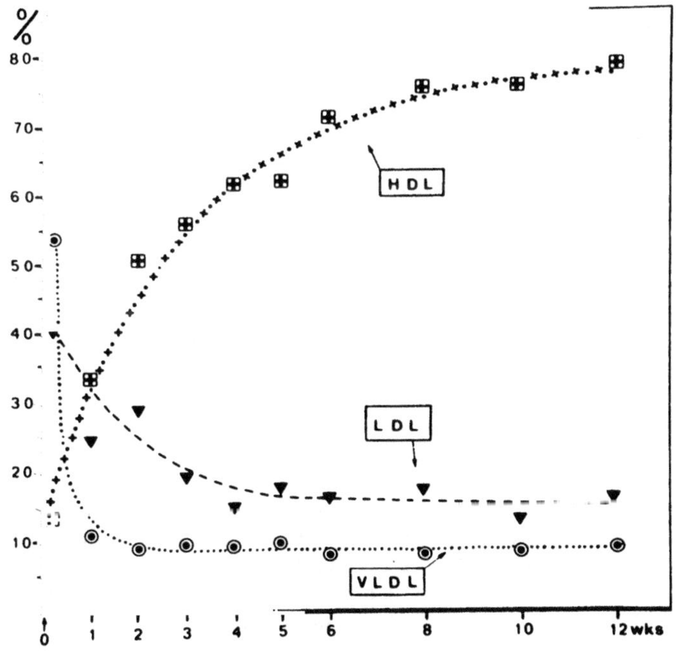

Figure 4: Progressive free cholesterol enrichment of HDL from cholesterol fed rabbits (33).

progressively enriched with free cholesterol (33) (Fig.4), possibly due to increased available substrate, without concomitant increase of lecithin cholesterol esterifying activity.

Lipoprotein modifications following cholesterol induction in rabbits are not restricted to the lipid composition, but include significant changes in the apoprotein moieties, particularly of d < 1.019 fractions. Shore et al. (34) indicated in particular, a marked increase of an apoprotein component, tipical of human type III hyperlipoproteinemia. This apoprotein, rich in arginine ("arg rich" or apo E) has been detected, as reported below, in several cholesterol-rich lipoprotein fractions, associated to the atherosclerotic process. D<1.019 lipoproteins from cholesterol fed rabbits display significant adhesiveness and easily show stacks, when examined by electron microscopy (35). These ultrastructural changes, as indicated above, may be reduced by corn-oil feeding (25).

Turnover experiments by Rodriguez et al. (36) on d<1.019 lipoproteins from control and cholesterol fed rabbits, demonstrated significantly reduced fractional catabolic rate (FCR) of ^{125}I labelled lipoproteins from cholesterolemic animals, and significantly increased uptake in the aortic wall, when these are injected into control animals. Analysis of radioactivity distribution, shows a markedly reduced conversion to lipoproteins of higher density. These findings are similar to those reported by Eisenberg et al. (37) for VLDL of type III individuals.

Guinea Pigs are particularly responsive to cholesterol administration. After feeding a diet containing 1% cholesterol, plasma, red cell and liver cholesterol levels are significantly elevated just within 1 week. After 10-12 weeks, tissue lipids and cholesterol increase several times above normal levels and a fatal hemolytic anemia usually develops (38).

Lipoprotein analysis following dietary induction shows significant changes both of LDL and HDL lipoproteins (Fig. 5) (39). LDL is markedly elevated in cholesterol fed, anemic guinea pigs and may be separated into two main fractions by gel chromatography. One fraction is similar to normal LDL, whereas to other is very rich in free cholesterol and appears as large transparent discs in electron microscopy. Stacked discs are also found

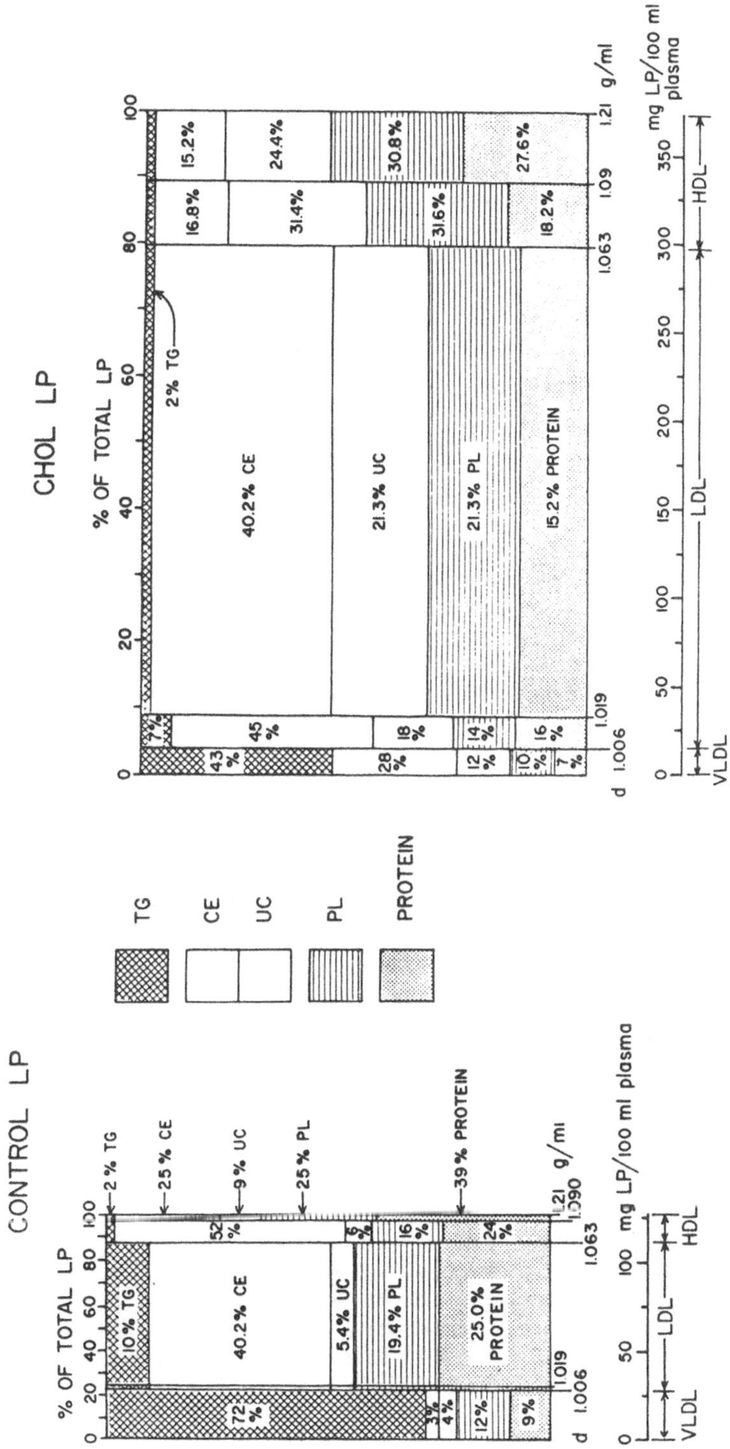

Figure 5: Lipoprotein changes following cholesterol induction in guinea pigs. A marked increase of LDL and HDL is noticeable (39).

in the HDL fraction. Apoprotein composition, both of LDL and HDL, is suggestive of the presence of apo E-like components, similarly to cholesterol fed rabbits.

Dogs are generally considered as refractory to atherosclerosis and poorly inducible by diet. When an atherogenic diet is however administered to thyroid-suppressed dogs, atherosclerosis develops in a large percentage of animals. Mahley et al. (40,41) analysed lipoprotein composition of hypothyroid dogs fed an atherogenic diet. These authors noted that "responders" and "non responders" could be readily identified among animals.

Whereas "non responders" have no atherosclerosis and a lipoprotein distribution and composition not markedly dissimilar from control animals, "responders" show high cholesterol concentrations in VLDL and the presence, in these, of an apoprotein component identifiable with apo E.

Miniature Swine - Spontaneous atherosclerosis of miniature swine can be accelerated by increased dietary cholesterol. Addition of lard and cholesterol to the diet of miniature swine 1-2 years of age markedly increases plasma total cholesterol levels (42). Lipoprotein fractionation demonstrates increased cholesterol in a lipoprotein fraction with α electrophoretic mobility and separable at density 1.063-1.087 (43). This cholesterol-ester rich HDL, designated HDL-C, has a high content of apoprotein E.

An interesting observation related to HDL-C is that this swine lipoprotein can inhibit HMG-CoA reductase, the key enzyme of cholesterol biosynthesis, when incubated with human fibroblasts (44). This finding shows that the so called "LDL-receptor" of human fibroblasts, documentedly absent in human homozygous type II hyperlipoproteinemia (45), can interact not only with apoprotein B, as generally believed, but also with other apoprotein components, i.e. probably apo E. Moreover, interactions with the "LDL-receptor" are not species specific, but may occur also with lipoproteins from different species.

Non Human Primates are a choice animal model for the induction of atherosclerosis, due to the documented similarities of their atherosclerotic lesions with those of man. Primates differ, however, remarkably in the dietary inducibility of hyperlipidemia and atherosclerosis (46). Some species are practically refractory to the dietary induction of hyperlipidemia and atherosclerosis, whereas others are very sensitive. In spite of the numerous studies on the atherogenic effects of different diets in primates, contributions on the lipoprotein changes following these diets are not abundant. Of particular interest, for most investigators, has been the analysis of lipoprotein similarities between primate and human hyperlipoproteinemias (46).

Rhesus monkeys have been the most thoroughly studied species. They are very sensitive to dietary cholesterol and develop atherosclerosis after several months of treatment. By monitoring lipoprotein distribution in Rhesus monkeys subjected to high dietary cholesterol, according to the Rudel procedure (47) with agarose gel--cromatography, it was possible to show significant alterations in the distribution of lipoproteins (Fig. 6). In particular, Macaca mulatta shows elevated plasma VLDL, IDL and LDL, with decreased HDL. LDL size is increased, as estimated from elution volumes; VLDL, markedly enriched with cholesterol esters, acquire β mobility on agarose electrophoresis. Therefore, lipoproteins of cholesterolemic Macaca have a distribution intermediate between that of human type II hyperlipoproteinemia and that of hypercholesterolemic rabbits.

A different pattern is that of the squirrel monkeys (Saimiris sciureus). After cholesterol induction, the agarose gel profile shows a striking increase of VLDL. LDL and HDL are not significantly changed. VLDL, similarly to rabbits, have increased cholesterol esters and electrophoretic β mobility. In line with the lipoprotein differences, the extent and distribution of atherosclerotic lesions is different between Macaca mulatta and Saimiris (48).

In the vervet monkeys (Cercopithecus ethiops) the pattern is still different. These animals are less responsive to the dietary induction of cholesterolemia. However, their chromatography elution profile indicates a dramatic increase of LDL, with slight reduction of HDL. LDL composition is not markedly modified. These monkeys are probably the most similar, in lipoprotein distribution, to the human type II hyperlipoproteinemia. The hypercholesterolemic response, maximal at 1 mg cholesterol/Kcal/day, is in the 300-400 mg/dl range, similar to that of the human disease. Distribution and morphology of atherosclerotic lesions have also been found to bear resem-

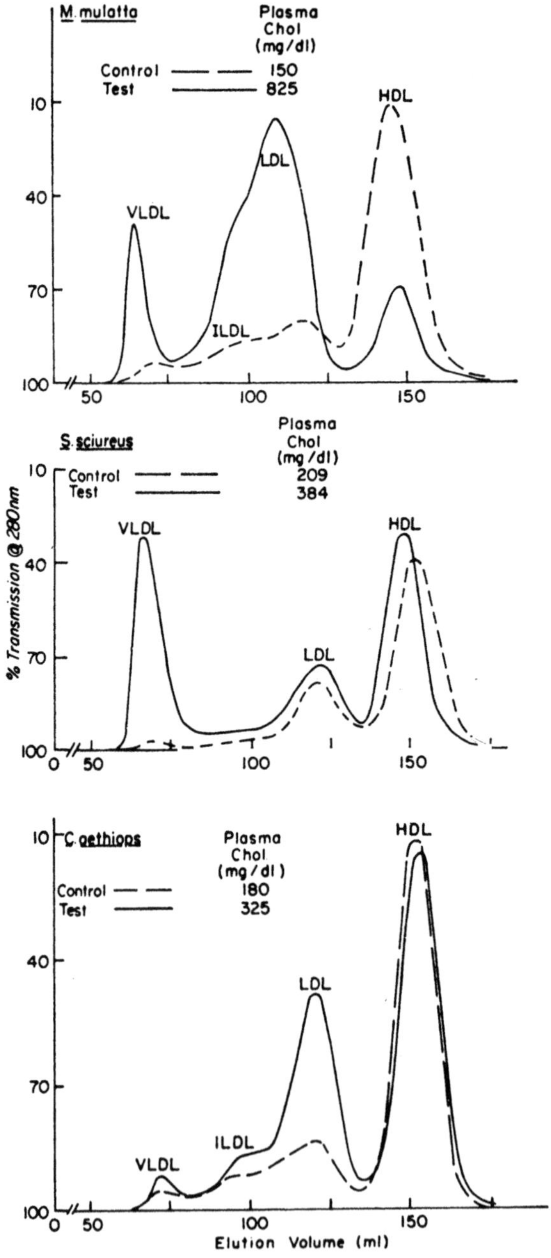

Figure 6: Changes of plasma cholesterol and lipoprotein distribution, analyzed by agarose gel chromatography, in primates fed atherogenic diets (46).

blance to those of human atherosclerosis (49).

Apoprotein composition of atherosclerotic monkeys has not been studied in great detail. It is generally assumed that LDL lipoproteins have a predominance of the B apoprotein and that an apo E may be found in cholesteremic VLDL. Interesting differences were however found in the HDL apoprotein and phospholipid composition of different primates, when compared to humans (50). It was in fact observed that, whereas chimpanzee has a phospholipid pattern in HDL, almost indentical to that of man (particularly rich in sphingomyelin and phosphatidylcholine, relatively low in phosphatidylinositol and phosphatidylethanolamine), baboon and Rhesus HDL are very high in phosphatidylcholine and relatively low in sphingomyelin as compared to humans and chimpanzee. Following dietaty induced hypercholesterolemia, chimpanzee, like man, has an increased sphingomyelin/phosphatidylcholine ratio. This is also found in Rhesus, but not in baboon. Chimpanzee has a typical hyper β-lipoproteinemia following cholesterol (it may have a hyper-pre β lipoproteinemia after sucrose and has a definitely decreased apo Al in both of these conditions) (50).

Turnover studies on primate lipoproteins have been described by Portman et al. (51). In squirrel monkeys, who, as pointed out, are similar to rabbits in their lipoprotein distribution following diet induced hypercholesterolemia, a slight reduction of LDL turnover is caused by hyperlipidemia, with increased absolute rates of synthesis and catabolism. In another study (52) the same Authors compared the effects of different dietary fats added to the cholesterol rich regimen. A higher fractional turnover rate of LDL was noted in monkeys when safflower oil was added to diets rich in cholesterol alone, butter + cholesterol, or butter alone. The fraction of the LDL pool remaining intravascular is also lower on diets containing safflower oil.

Avian Models- Birds, in particular pigeons and quails, respond with an elevation of plasma cholesterol levels following appropriate diets, cholesterol being carried predominantly in LDL (46).

They also display significant atherosclerotic lesions. The Japanese quail (Coturnix japonica) has elevated plasma cholesterol following a cholesterol-cholic acid rich diet; it also shows significant incidence of severe aortic and brachio-cephalic atherosclerosis. Chapman et al. (53) have described a selective breeding method for obtaining highly responsive quails.

Among pigeons, the White Carneau and Show Race breeds have a significantly different inducibility of atherosclerosis by diet (54). These two species differ in the oleic-linoleic acid ratio in aortic sterol esters. Comparison of responder and non responder pigeons has failed to describe specific differences in the cholesterol metabolism of these animals (55).

Rats - In spite of their extensive use for physiological studies on lipoprotein metabolism, rats have not been examined in detail for specific lipoprotein changes following atherogenic diets. Following addition of cholesterol to the diet, d < 1.006 lipoproteins are increased, and a new lipoprotein (d 1.006-1.030) may be detected, which contains immunologically identifiable HDL and LDL (56).

On cholesterol-cholic acid rich diets, added with propyl thiouracil, rats develop significant hypercholesterolemia and some atherosclerotic lesions. Breslow et al.(57) recently indicated that lipoproteins of d < 1.063, isolated from these rats, significantly suppress HMG-CoA reductase in rat liver cell cultures. This effect is not noted with lipoproteins from control animals.

Clinical Studies

In humans, addition of cholesterol to the diet, results in increased plasma cholesterol levels, predictable, in the general population, according to the formula

$$\Delta \text{ Chol} = 1.5 \sqrt{\Delta \text{ Chol}/1{,}000 \text{ KCal}}$$

suggested by Keys et al. (58). Analysis of individual responses shows, however, a considerable variability. Quintao et al. (59) in particular, showed that addition of cholesterol to the diet beyond certain limits fails to markedly increase plasma cholesterol levels. These Authors suggested that administration of very large cholesterol doses does not allow adequate absorption and, moreover, an extention of extra-vascular cholesterol pools may be at hand.

Recently, Nestel and Poyser (60) examined in detail

plasma cholesterol changes following two different daily intakes of cholesterol (250 and 750 mg). They observed that, of the eight tested patients, only three showed marked increases of cholesterolemia (all three had hypercholesterolemia), the other five having practically no change. On the whole, the mean cholesterol change of the total group was consistent with Keys' formula. Analysis of cholesterol absorption and excretion (both of acidic and neutral metabolites), indicated that, in the majority of patients, increased cholesterol intake is fully compensated by increased excretion, mostly under the form of a neutral sterols, and decreased synthesis. In some patients, cholesterol may be however retained, with a lesser change of neutral sterol excretion and little change of synthesis. According to the Authors, this data suggests that in some patients the cholesterol transport form may be changed to one less readily degraded.

Lewis (personal communication) has provided evidence that exogenously administered cholesterol may be transported in a VLDL fraction, thus giving rise to a form of type III hyperlipoproteinemia, similarly to the rabbit or to other animal species, as above outlined. This data awaits further confirmation, and may partly explain discrepancies in studies related to the dietary induction of hypercholesterolemia in man.

REDUCTION OF HYPERCHOLESTEROLEMIA

Diet induced hypercholesterolemia in animals may be readily reversed by changing the animal diet to the original type, thus achieving, at time, a regression of atherosclerotic lesions (61). In man, on the contrary, hypercholesterolemia is often of genetic origin and may be observed also on a "normal" diet, not particularly rich in cholesterol or saturated fat. Dietary treatment includes, therefore, reduction of dietary cholesterol and increased P/S ratio. Lipoprotein changes following increased P/S ratio have been reported in a preceding part of this paper. It is generally agreed that a high P/S ratio will cause a significant reduction of plasma cholesterol and particularly of the LDL fraction (62).

Classic dietary treatments reducing plasma cholesterol levels do not in general markedly affect lipoprotein composition. This has been well documented by Bagnall (63), who noted that in children with familial hypercholesterolemia, reduction of serum cholesterol fails to modify

LDL composition (Table 5). It may be assumed that cholesterol reducing diets will cause a drain of cholesterol, mostly from LDL and VLDL, not significantly modifying lipoprotein composition. Aside from the classic dietary treatments, new approaches have been recently described for the more severe forms of hypercholesterolemia. One of these is the intravenous hyperalimentation (64). This type of treatment, i.e. administration i.v. of solutions containing high percentages of dextrose and nitrogen, strikingly decreases plasma cholesterol levels also in patients with the homozygous form of type II hyperlipidemia. This approach is derived from the treatment of glycogen-storage disease, where hyperlipidemia is also reduced in this way, and which may find a definitive cure by portocaval shunt. I.v. hyperalimentation is now used as a presumptive test on the potential cholesterol lowering effect of portocaval shunt.

Following i.v. hyperalimentation marked decreases of plasma total and LDL cholesterol levels are noted. HDL cholesterol also decreases dramatically (Table 6). The relative composition of LDL after therapy is characterized only by a slight increase of triglycerides. LDL turnover in these patients is not significantly changed. The discussion on whether the mode of action of i.v. hyperalimentation is similar to that of portocaval shunt, has not brought to definitive conclusions. It is likely that the mechanism of i.v. hyperalimentation is more complex than, as suggested for portocaval shunt, elimination of the direct exposure of the liver to pancreatic hormones (65). The decrease of HDL is very significant and also hard to explain, but it is probably related to the mechanism of action of this special dietary treatment for hypercholesterolemia.

A simpler dietary approach for hypercholesterolemia has been recently suggested by Sirtori et al. (66), by substituting animal proteins in the diet with a soybean textured protein. Possibly due to aminoacid differences in the animal and vegetable proteins (67), plasma cholesterol levels are markedly decreased within two-three weeks in patients treated with the soybean regimen. Analysis of lipoprotein changes following this diet have shown that both LDL and VLDL cholesterol are significantly decreased (Table 7), whereas HDL cholesterol is practically unchanged. This data is related to 20 patients treated with soybean together with a diet with a high P/S ratio. Preliminary data on patients treated with the same regimen and low P/S ratio has shown that reduction of LDL

Table V

EFFECT OF DIET ON LDL COMPOSITION IN CHILDREN WITH FAMILIAL HYPERCHOLESTEROLEMIA (63)

	Normal	Type II	Diet
Chol.	197 ± 27	367 ± 55	265 ± 43
		% Composition of LDL	
CE	42.7 ± 2.7	45.2 ± 2.0†	45.1 ± 1.7†
FC	11.4 ± 1.4	12.7 ± 0.9†	12.2 ± 1.8
PL	21.5 ± 2.0	22.4 ± 2.3	23.3 ± 1.4
TG	6.0 ± 2.3	3.3 ± 1.0††	3.3 ± 1.9††
P	18.7 ± 2.1	16.4 ± 1.8	16.0 ± 1.9

† $P < 0.05$; †† $P < 0.01$ from Normals

Table VI

LIPID AND LIPOPROTEIN RESPONSE TO I.V. HYPERALIMENTATION (64)

		Total Chol.		mg/dl	LDL	Apo
		pre	post		pre	post
Patients	1	506	412		249	201
	2	545	409		272	207
	3	594	333		299	163
		HDL	Chol.	"	Triglycerides	
		pre	post		pre	post
Patients	1	21	11		61	70
	2	26	10		85	120
	3	26	17		133	115

Table VII

LIPOPROTEIN CHOLESTEROL CHANGES FOLLOWING THE SOYBEAN PROTEIN DIET (66)*

20 type II patients 3 weeks of soybean

CHOLESTEROL mg/dl	Pre	Post	Δ
$\bar{X}$ VLDL	52	31	21^{+}
$\bar{X}$ LDL	233	186	47^{++}
$\bar{X}$ HDL	40	39	1
TOTAL	325	256	69^{++}
TRIGLYCERIDES	197	165	32^{+}

$^{+}P < 0.05$; $^{++}P < 0.01$

*mean levels of 20 patients treated with the soybean protein diet for three weeks, either before or after a low-lipid low-cholesterol diet.

cholesterol is less significant in these, whereas total cholesterolemia is decreased not differently than with a high P/S ratio. More detailed studies on lipoprotein composition in these patients are being carried out.

CONCLUSIONS

Dietary changes may modify lipoprotein composition, in a few cases dramatically. Examples of these are the animal models of hypercholesterolemia, of ethanol or CHO induced hypertriglyceridemias in humans, and of intravenous hyperalimentation in hypercholesterolemic patients. In other cases, modifications may be slight or hardly significant. Detailed compositional studies of plasma lipoproteins are awaited to exactly determine site and mode of action of the different diets.

REFERENCES

1. Quarfordt, S.H., Frank, A., Shames, D.M., Berman, M. and D. Steinberg. 1970. Very low density lipoprotein triglyceride transport in Type IV hyperlipoproteinemia and the effects of carbohydrate-rich diets. J. Clin. Invest 49: 2281-2297

2. Wilson, D.E., Schreibman, P.H., Breston, A.C.,and R.A. Arky. 1970. The enhancement of alimentary lipemia by ethanol in man. J. Lab. Clin. Med. 75: 264-273.

3. Schonfeld, Y.1970. Changes in the composition of very low density lipoprotein during carbohydrate induction in man. J. Lab. Clin. Med. 75: 206-211.

4. Ruderman, N., Jones, A.L., Krauss, R., and E. Shafrir. 1971. A biochemical and morphologic study of very low density lipoproteins in carbohydrate induced hypertriglyceridemia. J. Clin. Invest. 50: 1355-1368

5. Carlson, L.A., and D. Ballantyne. 1976. Changing relative proportions of apolipoproteins CII and CIII of very low density lipoproteins in hypertriglyceridemia. Atherosclerosis 23: 563-568.

6. Havel, R.J., Fielding, C.J., Olivecrona, T., Shore V.G., Fielding P.E., and T. Egelrud. 1973. Cofactor activity of protein components of human very low density lipoproteins with hydrolysis of triglycerides by lipoprotein lipase from different sources. Biochemistry 12: 1828-1832

7. Schonfeld, G., Weidman, S.W., Witztum, J.L., and R.M. Bowen. 1976. Alterations in levels and interrelations of plasma apolipoproteins induced by diet. Metabolism 25: 261-275.

8. Mendelson, J.H., and N.K. Mello. 1973. Alcohol-induced hyperlipidemia and beta lipoproteins. Science 180: 1372-1374

9. Puglisi, L. Caruso, V., Conti, F., Fumagalli, R., and C. Sirtori. 1977. Effect of hypolipidemic and hypoglycemic drugs on ethanol induced hypertriglyceridemia in rats. Pharm. Res. Comm. 9: 71-77

10. Losowsky, W.S., Jones, D.P., Davidson, C.S., and C.S. Lieber. 1963. Studies of alcoholic hyperlipidemia and its mechanism. Am. J. Med. 35: 794-803.

11. Avogaro, P., and G. Cazzolato. 1975. Changes in the composition physico-chemical characteristics of serum

lipoproteins during ethanol-induced lipaemia in alcoholic subjects. Metabolism 24: 1231-1242.

12. Avogaro P. 1976. Ethanol-lipaemia , a doubtful risk factor. Proc. IV Int. Symposium on Atherosclerosis. Tokyo, in press.

13. Sirtori, C.R., Biasi, G., Vercellio, G., Agradi, E., and E. Malan. 1975. Diet, lipids and lipoproteins in patients with peripheral vascular disease. Am. J. Med. Sci. 268: 325-332.

14. Miller, G.J. and N.E. Miller. 1975. Plasma high-density lipoprotein concentration and development of ischaemic heart - disease. Lancet i: 16-19.

15. Levy, R.I., Bonnell, M., and N.D. Ernst. 1971. Dietary management of hyperlipoproteinemia. J. Am. Ass. 58:406-416.

16. Wilson D.E. and R.S. Lees. 1972. Metabolic relationships among the plasma lipoproteins. J. Clin. Invest. 51: 1051-1057.

17. Carlson, L.A., Olsson, A.G., Orö, L., Rössner, S., and G. Walldius. 1974. Effects of hypolipidemic regimens on serum lipoproteins. Proc. III Int. Symposium on Atherosclerosis, Berlin, 768-781.

18. Eisenberg, S. 1976. Lipoprotein metabolism and hyperlipemia. In: Atherosclerosis Reviews Vol. 1 Paoletti R., Gotto A.M., Eds. Raven Press. N.Y., 23-60.

19. Vessby, B., and M. Lithell.1976. Dietary effects on lipoprotein levels in hyperlipoproteinemia. Delineation of two subgroups of endogenous hypertriglyceridemia. Artery 1: 63-85

20. Belin, J., Smith, A.D., and R.H.S. Thompson. 1975. Effect of short-term oral administration of sunflower seed oil on the pattern of non-esterified fatty acids in human plasma. Clin. Chim. Acta 61: 95-98.

21. Kingsbury, K.J., Brett, C., Stovold, R., Chapman, A., Anderson, J., and D.M. Morgan. 1974. Abnormal fatty acid composition of human atherosclerosis. Postgrad. Med. J. 50: 425-440

22. Kingsbury, K.J., Morgan, D.M., Stovold, R., Brett, G.G., and J. Anderson. 1969. Polyunsaturated fatty acids and myocardial infarction. Lancet ii: 1325--1331.

23. Ockner, R.K., Hughes, F.B. and K.J., Isselbacher. 1969. Very low density lipoproteins in intestinal

lymph: role in triglyceride and cholesterol transport during fat absorption. J. Clin. Invest. 48: 2367-2373.

24. Spritz, N. and M.A. Mischkel. 1969. Effects of dietary fats on plasma lipids and lipoproteins: an hypothesis for the lipid-lowering effect of unsaturated fatty acids. J. Clin. Invest. 48: 78-86.

25. Stange, E., Agostini, B., and J. Papenberg. 1975. Changes in rabbit lipoprotein properties by dietary cholesterol, and saturated and polyunsaturated fats. Atherosclerosis 22: 125-148.

26. Rodriguez, J.L., Catapano, A., Ghiselli, G.C., and C.R. Sirtori. 1976. Very low density lipoproteins in normal and cholesterol-fed rabbits: lipid and protein composition and metabolism. Part. 1. Chemical composition of very low density lipoproteins in rabbits. Atherosclerosis 23: 73-83.

27. Jackson, R.L. and A.M. Gotto jr. 1974. Phospholipids in biology and medicine (second of two parts). N. Engl. J. Med. 290: 87-93.

28. Chait, A., Onitiri, A., Nicoll, A., Rabaya, E., Davies, J., and B. Lewis. 1974. Reduction of serum triglyceride levels by polyunsaturated fat. Studies on the mode of action and on very low density lipoprotein composition. Atherosclerosis 20: 347-364.

29. Taunton, O.D., Morrisett, J.D., Segura, R., Pownall, M.J., Jackson, R.L., and A.M. Gotto. 1974. Effect of dietary fatty acid composition on lipoprotein structure. Circulation 49-50, Suppl. III, Abs. 1050

30. Anitschkow, N.N. 1913. Über die Veränderung der Kaninchen aorta bei experimenteller Cholesterinsteatose. Beitr. path. Anat. 56: 379-383.

31. Shore, B., and V. Shore. 1976. Rabbits as a model for the study of hyperlipoproteinemia and atherosclerosis. Adv. Exp. Med. Biol. 67:123-141.

32. Brattsand, R. 1976. Distribution of cholesterol and triglycerides among lipoprotein fractions in fat-fed rabbits at different levels of serum cholesterol. Atherosclerosis 23: 97-110, 1976.

33. Pinon, J.C., and A.M. Bridoux. 1977. High density lipoproteins in cholesterol fed-rabbits: progressive enrichment with free cholesterol. Artery 3: 59-71.

34. Shore, V.G., Shore, B., and R.G. Hart. 1974. Changes in apolipoproteins and properties of rabbit very low density lipoproteins on induction of cholesterolemia.

Biochemistry 13: 1579-1582.

35. Agostini, B., Seidel, D. and M. Wieland. 1973. Electron microscopy of low-density plasma lipoproteins in patients with Type IIa hyperlipoproteinemia. Naturwissenschaften 60: 111-113.

36. Rodriguez, J.L., Catapano, A., Ghiselli, G.C., and C. Sirtori. 1976. Very low density lipoproteins in normal and cholesterol-fed rabbits: lipid and protein composition and metabolism. Part 2. Metabolism of very low density lipoproteins in rabbits. Atherosclerosis 23:85-96

37. Bilheimer, D.W., Eisenberg, S. and R.I. Levy. 1971. Abnormal metabolism of very low density lipoproteins (VLDL) in Type III hyperlipoproteinemia. Circulation 44: Suppl. II, Abs. 186.

38. Green, M.H., Crim, M., Traber, M., and R. Ostwald. 1976. Cholesterol turnover and tissue distribution in the Guinea Pig in response to dietary cholesterol. Nutrition 106: 516-528.

39. Sardet, C., Hansma, H., and R. Ostwald. 1972. Characterization of guinea pig plasma lipoproteins: the appearance of new lipoproteins in response to dietary cholesterol. J. Lipid. Res. 13:624-639.

40. Mahley, R.W., and K.H. Weisgraber. 1974. Canine lipoproteins and atherosclerosis I. Isolation and characterization of plasma lipoproteins from control dog. Circ. Res. 35: 713-721.

41. Mahley, R.W., Weisgraber, K.H., and I. Innerarity. 1974. Canine lipoproteins and atherosclerosis II. Characterization of the plasma lipoproteins associated with atherogenic and non-atherogenic hyperlipidemia. Circ. Res. 35: 722-733.

42. Florentin, R.A., and S.C Nam. 1968. Dietary-induced atherosclerosis in miniature swine. I. Gross and light microscopy observation: Time of development and morphologic characteristics of lesions. Exp. Mol. Pathol. 8: 263-301.

43. Mahley, W.R., Weisgraber, K.H., Innerarity, T., Brever, H.B., and G. Assman. 1975. Swine lipoproteins and atherosclerosis. Changes in the plasma lipoproteins and apoproteins induced by cholesterol feeding. Biochemistry 14: 2817-2823.

44. Bersot, T.P., Mahley, R.W., Brown, M.S., and J.L. Goldstein. 1976. Interaction of swine lipoproteins

with the low density lipoprotein receptor in human fibroblasts. J. Biol. Chem. 251: 2395-2398.

45. Goldstein, J.L., and M.S. Brown. 1973. Familial hypercholesterolemia: identification of a defect in the regulation of 3-hydroxy-3-methylglutaryl-coenzime A reductase activity associated with overproduction of cholesterol. Proc. Nat. Acad. Sci. 70: 2804-2808

46. Clarkson, T.B., Prichard, R.W., Bullock, B.C., St. Clair, R.W., Lehner, N.D.M., Jones, D.C., Wagner, W.D., and L.L. Rudel. 1976. Pathogenesis of atherosclerosis: some advances from using animal models. Exp. Mol. Path. 24:264-286.

47. Rudel, L.L., Lee, J.A., Morris, M.D., and J.M. Felts. 1974. Characterization of plasma lipoproteins separated and purified by agarose-column chromatography. Biochem. J. 139: 89-95.

48. Clarkson, T.B., Prichard, R.W., Bullock, B.C., Lehner, N.D.M., Lofland, H.B., and **R.W. St. Clair**. 1970. Animal models of atherosclerosis. Animal models. Biomed. Res. 3: 22-41.

49. Bullock, B.C., Lehner, N.D.M., Clarkson, T.B., Feldner, M.A., Wagner, W.D., and H.B. Lofland. 1975. Comparative primate atherosclerosis I: Tissue cholesterol concentration and pathologic anatomy. Exp. Mol. Pathol. 22: 151-175.

50. Blaton, V., and H. Peeters. 1976. The non-human primates as models for studying human atherosclerosis: studies on the Chimpanzee, the Baboon and the Rhesus Macacus. Adv. Exp. Med. Biol. 67:33-64.

51. Portman, O., Illingworth, D.R., and M. Alexander. 1975. The effects of hyperlipidemia on lipoprotein metabolism in Squirrel monkeys and rabbits. Biochim. Biophys. Acta. 398: 55-71.

52. Portman, O., Alexander, M., Tanaka, N., and P. Soltys. 1976. The effects of dietary fat and cholesterol in the metabolims of plasma low density lipoprotein apoproteins in Squirrel monkeys. Biochim. Biophys. Acta. 450:185-196.

53. Chapman, K.P., Stafford, W.W., and C.E. Day. 1976. Animal model for experimental atherosclerosis produced by selective breeding of japanese quail. Adv. Exp. Med. Biol. 67:347-356.

54. Ravi Subbiah, M.T., Kottke, B.A., and I.A. Carlo. 1974. Studies in spontaneously atherosclerosis

susceptible and resistant pigeons: nature of plasma and aortic sterol, steryl esters, and free fatty acids. Int. J. Biochem. 5, 63-68.

55. Wagner, W.D., and T. B. Clarkson. 1974. Mechanisms of the genetic control of plasma cholesterol in selected lines of Show Racer pigeons. Proc. Soc. Exp. Biol. Med. 145:1050-1057.

56. Lasser, N.L., Roheim, P.S., Edelstein, D. and H.A. Eder. 1973. Serum lipoproteins of normal and cholesterol-fed rats. J. Lipid. Res. 14: 1-8.

57. Breslow, J.L., Spaulding, D.R., Lothrop, D.A., and A. Clowes. 1975. Effect of lipoprotein on 3-hydroxy--3-methylglutaryl coenzyme A (HMG CoA) reductase activity in rat liver cell culture: special suppressant effect of a lipoprotein isolated from hypercholesterolemic rat plasma. Biochem. Biophys. Acta. 67: 119-125.

58. Keys, A., Anderson, J.T., and F. Grande. 1965. Serum cholesterol response to changes in the diet II. The effect of cholesterol in the diet. Metabolism 14: 759-765.

59. Quintao, E., Grundy, S.M., and E.H. Ahrens jr. 1971. Effects of dietary cholesterol on the regulation of total body cholesterol in man. J. Lipid. Res. 12: 233-247.

60. Nestel, P.J., and A. Poyser. 1976. Changes in cholesterol synthesis and excretion when cholesterol intake is increased. Metabolism 25: 1591-1599.

61. Armstrong, M.L., 1976. Regression of atherosclerosis. In: Atherosclerosis Reviews Vol. 1. Paoletti R., Gotto A.M., Eds. Raven Press. N.Y., 137-182.

62. Levy, R.I. 1972. Dietary and drug treatment of primary hyperlipoproteinemia. Ann. Int. Med. 77: 267-294.

63. Bagnall, T.F. 1972. Composition of low density lipoprotein in children with familial hyperbetalipoproteinemia and the effect of treatment. Clin. Chim. Acta. 42:229-233.

64. Torsoik, H., Feldman, H.A., Fischer, J.E., and R.S. Lees. 1975. Effects of intravenous hyperalimentation on plasma-lipoproteins in severe familial hypercholesterolaemia. Lancet i: 601-607.

65. Ahrens, E.H. jr. 1974. Homozygous hypercholesterolemia and the portacaval shunt. Lancet. ii, 449-451.

66. Sirtori, C.R., Agradi, E., Conti, F., Mantero, O.,

and E. Gatti. 1977. Soybean-protein diet in the treatment of type II hyperlipoproteinemia. Lancet i: 275-278.

67. Carroll, K.K., and R.M.G. Hamilton. 1975. Effects of dietary protein and carbohydrate on plasma cholesterol level in relation to atherosclerosis. J. Food Sci. 40: 18-23.

Conclusion

LIPOPROTEIN MOLECULE OR LIPOPROTEIN PARTICLE?

Hubert Peeters

Lipid and Protein Department, LBS

B-1180 Brussels, Belgium

May I wind up this Advanced Study Institute by sketching briefly what this course has meant at least as far as my judgement goes. Not that I believe to be a polyvalent expert on all topics presented but I do think it is my duty to outline the status of our knowledge and the avenues along which we should make progress. What do we expect to happen, what is to become important, what should be looked for?

The first line of thought is the apoprotein line. This line has been complex but rewarding and has been shedding light on a set of clinical problems nobody hoped to solve a few years ago. The beautiful thing about apoproteins is their potential in the interpregation of the ultracentrifugal and electrophoretic protein patterns. Two physical parameters - gravitational density during centrifugation and optical density after electrophoresis - were in need of an objectivation through the primary structure of the apopolypeptides. The distribution of the apopolypeptides interprets quite a few problems of lipoprotein density classes but does not solve the entire problem. If only the exchange of lipids during centrifugation under the effect of high salt concentrations does call for further careful and critical assessment of UCF data. In my view the recentrifugation of electrophoretic lipoprotein fractions is also a must. With modern knowledge on apoproteins and their quantitation, completed by a careful and complete lipid analysis such as of the phospholipids, a profound and motivated comparative countercheck of ultracentrifugal and electrophoretic fractions should be performed. Several times in the course of these days important physiological shifts inside density classes were documented. This confirms my opinion that still more comparative work is required.

Systematics in lipoproteins will continue to fulfill a unique role: they absorb and integrate knowledge from an increasing number of disciplines and in turn generate important data which are useful to these other disciplines by identifying new species of molecules and by shedding light on their interrelationships.

We touched on two problems: apoprotein quantitation and lipid quantitation. We should accept that neither one of these problems is really solved. Elegant and efficient methods can be expected and are indeed needed to rapidly and precisely quantitate proteins and lipids. Where is the comprehensive paper on delipidation? What is the comprehensive task of apoprotein? Most problems with lipoproteins originate from their mysterious amphiphilic properties. All our analytical methods struggle with two antithetic requirements: to isolate valid protein and valid lipid out of a molecule which has been built to swim like a fish in water. A dry duck is a lame duck: an apoprotein seems hard to dissolve. On the contrary a lipoprotein is beautifully stabile in our aqueous biological fluids.

As a consequence of this problem the importance of physical methods which can look at the intact molecule becomes evident and this is a field where more progress is to occur. In the past centrifugation and electrophoresis did more for lipoproteins than all cholesterol determinations. And cholesterol anyhow became truly interesting only when related to alpha or beta protein, thus when related to lipoprotein. Here then is an incidental remark: we need to introduce molar expression of lipids and in turn express these in function of lipoproteins: if we do not succeed in doing this the clinicians will persevere for a long time in a rather rough approximation of dyslipaemia as a kind of sophisticated form of hypercholesterolaemia whereas the core of the matter lies in a profound and disturbed phospholipid-apoprotein relationship followed or preceded by triglyceride and cholesterol ester disorders. If we group these remarks the need for estimations and parameter evaluations of intact lipoproteins is a corollary as well as a prerequisite. A method such as microviscosity may be a step in the good direction.

Which of these questions is the prime mover? All approaches to interpretation of the metabolic pathways of lipoproteins are still blocked by a set of serious questions:

- Site and rate of synthesis of apoproteins
- Mechanisms of lipidation after apoprotein synthesis
- Role of succession of incorporation of lipid classes
- Interexchanges of lipids between secreted lipoproteins
- Catabolism of lipoprotein lipids
- Relipidation and reuse of remnant apoproteins after their delipidation in vivo.

One could organize quite a few round tables around such topics.

We shall go home with such questions unanswered.

To my mind any physical method aimed at the structura interpretation of the lipoprotein molecule is due to allow for a fundamental interpretation of lipoprotein metabolism. The metabolic problems will not be solved through metabolic methods alone, the metabolic problems require a profound knowledge of the qualities of the lipoprotein molecule itself. Physical parameters can be much shorter to obtain than a panel of chemical data.

Evidently comparative biochemistry including apoprotein and lipid analysis cannot be overlooked. Such studies are easier to perform than the work on the exceedingly rare congenital disorders in man. Animals will become an excellent tool when the knowledge of the biophysical relationship between protein and lipid will be better understood because extrapolation, for instance of data on drug effects, will thus become feasible. This is not yet possible today.

Is the word lipoprotein molecule correct? I guess not: we should have discussed the lipoprotein particle and this criticism of the title of the Institute itself, is proof of our awareness of our ignorance. To attack lipoproteins - molecules or particles - also requires the help of immunology. I touched on this problem indirectly when I mentioned the quantitation problem, but immunology of lipoproteins indeed has its home-made problems: antisera see antigens when they are naked. Do the lipids cover up antigenic sites? How often? How much? These remarks on immunology bring us into the heart of the matter: lipoproteins are not "beautiful people" for a few selected scientists. They actually hurt the vessel wall and this kills people, including the beautiful. So the natural history, the pathology, and the search for drugs and diets are of the utmost importance. This part of the message, though stated very briefly, should stay with us.

What do I suggest to the new students in the field?

a. To read the older papers. Several times during these days facts have been rediscovered which most of us had forgotten.
b. To remark that people die in spite of our perfectly equipped laboratories.
c. To ask for more and for better laboratories. Scramble for new physical and clinical methods as well for apoprotein as for lipid analysis. Make comparative studies so that your progress can be used over a broad field of research in the lipoprotein field to which we belong.
d. To establish contact not only with your kinsmen but with people of different disciplines in the field of dyslipaemia. The same sample can be looked at by several people.
e. To exchange methods, standards, procedures on the practical level. We have been deceptively poor in this respect, I know

this. At least bridges were built, contacts were established.

A second set of propositions covers a set of topics in which I believe:

1. Apoprotein quantitation as a clinical tool
2. Phospholipid distribution in lipoproteins
3. Link plasma, platelet, and arterial wall lipids, especially phospholipids
4. Look for acyl transfer mechanisms requiring other phospholipids than lecithin as a substrate
5. If you have a physical method available, become an expert, stick to your guns. Use precise polypeptides and serial systematic lipids preferably the synthetic and semi-synthetic ones. Do not use a physical method on a lipoprotein until you know what exactly happened with a model system.
6. Fatty acids were not mentioned till this morning. Reread then the old papers, reinterpret them and repeat the old clear cut experiments that withstood the tug of time, and look now for fatty acid movements inside lipoprotein classes, and inside their lipids.
7. Use experimental animals only if you have access to polypeptide and lipid data. Do not feed rabbits, SEA quail, or pigeons, rapidly measure their cholesterol and write a paper. You cannot correctly assess nor evaluate changes in lipids if you cannot relate them to apoprotein content.
8. Participate in any drug trial they propose to you but only if the company accepts your minimal clinical and laboratory guidelines as referred to in the previous point.

What do we expect to achieve ourselves?

- To develop SM/PC ratio as a clinical concept (sphingomyelin/lecithin)
- To develop microviscosity as a physical measurement of lipoprotein
- To discover carrier proteins for phospholipids. They should act as messengers at the site of lipoprotein synthesis. They have to be specific for each phospholipid and sensitive to the fatty acid load of the phospholipid. If you consider the chronobiology of cholestereol synthesis in the rat for instance, such rythmic succession of specific enzyme and carrier proteins will be requested.

The informational potential of a lipoprotein is tremendous: if we persevere we shall break its code for the benefit of mankind and just a little to our own satisfaction.

INDEX

GPSR Compliance

The European Union's (EU) General Product Safety Regulation (GPSR) is a set of rules that requires consumer products to be safe and our obligations to ensure this.

If you have any concerns about our products, you can contact us on ProductSafety@springernature.com

In case Publisher is established outside the EU, the EU authorized representative is:

Springer Nature Customer Service Center GmbH
Europaplatz 3
69115 Heidelberg, Germany

Batch number: 10615115

Printed by Printforce, the Netherlands